# THE COAST OF
# ENGLAND AND WALES
# IN PICTURES

# THE COAST OF
# ENGLAND AND
# WALES
## IN PICTURES

WITH A COMMENTARY BY
J. A. STEERS

CAMBRIDGE
AT THE UNIVERSITY PRESS
1960

PUBLISHED BY
THE SYNDICS OF THE CAMBRIDGE UNIVERSITY PRESS

Bentley House, 200 Euston Road, London, N.W.1
American Branch: 32 East 57th Street, New York 22, N.Y.

THIS EDITION

©

CAMBRIDGE UNIVERSITY PRESS
1960

First Edition (*A Picture Book of the Whole Coast of
England and Wales*)                1948
This Edition                1960

*Printed in Great Britain at the University Press, Cambridge
(Brooke Crutchley, University Printer)*

# CONTENTS

# PREFACE

Shortly after the publication of *The Coastline of England and Wales* (1946), Mr Kendon of the Cambridge University Press suggested that the pictures in that book, prefaced by an introduction by Dr Fraser Darling and an essay by myself, should form a separate volume which was eventually published in 1948 under the title of *A Picture Book of the Whole Coast of England and Wales*. This was not reprinted.

In 1958 Mr Kingsford discussed with me the possibility of either a new edition of the *Picture Book*, or some modification of it. The appearance of the photographs in 1948 was not always successful, partly because some of the pictures themselves were not satisfactory, and partly because the paper available at that time did not favour first-class reproduction. Since then the University, through the skill of Dr J. K. S. St Joseph, has accumulated a magnificent collection of air photographs, many of which relate to the coast. The juxtaposition of new air and old ground photographs would have been unsatisfactory. Hence I have gone back to the originals of the ground photographs, and all the illustrations have been reproduced by a different method.

A judicious selection of air and ground photographs probably gives the best impression of the coast, since it enables the reader to combine general and particular views in a way not possible if all are taken either from the air or on the ground. It would be only too easy to increase the number of pictures, but the cost of production sets strict limits, and makes it impossible to add, for example, views of seaside resorts if—and this is the primary aim of the book—as complete a coverage of the natural coast as possible is attempted.

The reader of the book must imagine that he is making a journey from London via the Channel coast to Land's End, up the Bristol Channel to near Cardiff, and then along the coast of South Wales to the far end of the Tenby peninsula. The route then follows the Pembrokeshire coast, Cardigan Bay, and the Lleyn Peninsula to the Menai Strait. A brief reconnaissance of Anglesey precedes the continuation along the coast of North Wales, Lancashire, and Cumberland to the Solway Firth. The journey recommences at Berwick and follows the east coast as far as the Thames.

In order to make a reasonable, although short, commentary on important coastal features, the photographs are arranged in groups, each of which refers to a section of the coast. The limits chosen for the sections are quite arbitrary. A few text-figures and maps have been included to render the descriptions more intelligible.

Once again I have to thank my old friend Dr H. Dighton Thomas for his kindness in eliminating errors and making helpful suggestions when this book was in typescript. To Dr St Joseph I am particularly grateful for his photographs and also for helping me to give titles to them.

CAMBRIDGE                                                                                                    J. A. STEERS

# ACKNOWLEDGEMENTS

The following list gives the names of the photographers and agencies responsible for the photographs, and to whom the publishers are grateful for permission to reproduce their work:

Aerofilms Ltd, London: 8, 26, 28, 96, 114, 142.

Aero Pictorial Ltd, Redhill, Surrey: 37, 42.

The Air Ministry (Crown Copyright reserved): 161 (and see below).

Hallam Ashley, Norwich: 154, 157.

Central Office of Information, the Dixon-Scott Collection (Crown Copyright reserved): 4, 11, 18, 41, 43, 45, 48, 53, 66, 92, 94, 98, 122, 135, 139, 155.

James Gibson, St Mary's, Isles of Scilly: 49, 50.

Her Majesty's Stationery Office; photographs from the Geological Survey (Crown Copyright reserved): 2, 7, 9, 10, 20, 36, 39, 46, 47, 59, 60, 62, 63, 64, 67, 69, 75, 77, 78, 84, 85, 89, 97, 109, 115, 116, 117, 118, 120, 121, 124, 125, 127, 128, 130, 132, 133, 134, 136, 138, 143, 150, 164.

Eric Kay, Ringwood, Hants: 23, 27, 29, 31, 32, 34.

J. A. Lofthouse: 51.

Photographs by J. K. S. St Joseph, reproduced by permission of the Air Ministry (Crown Copyright), and the Department of Aerial Photography, University of Cambridge: 1, 3, 5, 6, 12, 13, 14, 15, 16, 17, 19, 21, 22, 24, 25, 33, 35, 38, 40, 44, 52, 54, 55, 56, 57, 58, 61, 65, 68, 70, 71, 72, 73, 74, 76, 79, 80, 81, 82, 83, 86, 87, 88, 90, 91, 93, 95, 99, 100, 101, 102, 105, 106, 107, 108, 110, 112, 113, 119, 123, 126, 129, 131, 140, 141, 144, 145, 148, 149, 151, 152, 153, 156, 158, 159, 160, 162, 163, 165, 166, 167.

J. A. Steers: 103, 104, 146.

H. H. Swinnerton: 147.

*The Times*: 30, 111.

Messrs Raphael Tuck and Will F. Taylor: 137.

# INTRODUCTION

# *The Changing Coast*

In continental stretches like America, Russia, or Australia and, of course, in many smaller regions it is possible to travel far without noting any great change in scenery. But the British Isles possess both a very wide variety of scenery within a small area and also examples of rocks formed in nearly all the geological ages through which the earth has passed. The variety of rocks in England and Wales is largely responsible for the many types of coastline in the country.

A recent estimate of the length of the coastline of England and Wales is 2750 miles, and it is very rare to find the same kind of coastal scenery for more than ten or fifteen miles together.

## CLIFFS

Although it is true that a particular *type* of coast may extend for a long distance, yet the details of the rocks vary from place to place. In Cornwall the granite cliffs of Land's End with their castellated features (**53**) stand in marked contrast to the more sloping forms of the highly contorted rock-formation near Bude (**93**) or to the cliffs of Lantivet Bay (**43**). Examples can also be found in other counties; in the variation in form between the flat-topped and the 'hog's-back' cliffs of North Devon (**61, 68**) and in the contrast between the highly coloured Old Red Sandstone on the southern side of St Bride's Bay with the beautiful cliffs of Cambrian rock near St David's and Solva (**89** *a, b*), or again the difference between the part of Dorset near Lyme Regis (**32**) and that nearer Lulworth (**26, 28**) is so striking that we are apt to underrate its significance.

The forms of cliffs depend largely on a few main factors. These include the rate at which the sea cuts into the foot of the cliff in relation to the rate at which atmospheric processes or weathering wear back the higher parts, the initial slope of the land (compare **47** and **130**), the dip of the rocks (**76, 97, 150**), and the effect of the land water draining out of the rocks in small gullies (**10**), which, although at first insignificant, may by the process of cutting back develop into marked re-entrants which are further hollowed out by the action of the waves.

Moreover, the composition of the rocks, whether they are hard or soft, resistant or easily worn away, matters considerably. Along the coast of Holderness (**144**) the encroachment of the sea is as much as five or six feet a year. The cliffs are of boulder-clay, a substance easily

I

eroded or washed away by the constant action of the waves. It has been estimated with some accuracy that the total loss of land on the stretch of coast between Bridlington and Spurn Head since Roman times amounts to about 80 square miles. Highly important also is the presence of any clayey water-logged beds which, if inclined seaward, facilitate the slipping of the over-lying beds. The Folkestone and Dowland Cliffs (5, 33) illustrate this point and contrast strongly with the vertical chalk cliffs near Birling (11).

Another factor of consequence in cliff-form depends upon recent (in the geological sense) vertical movements of land in relation to sea-level, on account of which rock platforms which formerly were at sea-level or below are now visible well above that level. Where these rock platforms or raised beaches occur (46), the cliff above them is clearly not entirely caused by present-day conditions. A walk along the Gower coast shows this very well: there are many stretches of the former beach which remain at the foot of the cliffs, thus making the present cliff outline composite.

Many of the more spectacular features of cliffs are connected with either faults or joints in the rocks. These are lines of weakness which offer less resistance to the elements, marine and atmospheric, so that in places great gashes may be cut, for example the gully known as the Huntsman's Leap (84) not far from Tenby. On a smaller scale the effect of joints in igneous rock is well seen at Land's End (53), and in the Whin Sill of Northumberland where it runs out at Dunstanburgh Castle (129) and Cullernose Point (130) (see also Fig. 11). Caves, too, are often hollowed out along joints and other planes of weakness (132, 142). Sometimes blow-holes are formed (143).

In addition to all these factors, the original form of the land plays a predominant part. At some time or other the sea came to rest against a land-mass, of which the detailed shape is for the most part a matter of conjecture. The present cliffs may be the result of the waves washing constantly at the foot of the original slopes and nothing more; the hog's-back cliffs of North Devon and the soft cliffs of Norfolk and Suffolk illustrate this. Inside the wide sand formation known as Morfa Harlech (103) there is a great range of hills the base of which must at one time have been washed by the waves. It would, nevertheless, be rash to assume that their slopes have been to any extent changed by wave-action, because the sheltered nature of this part of Cardigan Bay has to be considered. In many parts of Cornwall, too, it is not by any means easy to say how far the erosive action of the sea is responsible for the present shape of the cliffs (44, 45, 47, 56, 61).

Along many stretches of coast the cliffs are fronted by a lower or undercliff. This is usually the result of landslips, but may be closely related to the formation of the raised beaches (p. 4). There are two particularly well-known coastal landslips, Folkestone Warren and Dowlands near Lyme Regis. At Folkestone Warren (5) a crumpled mass lies at the foot of the chalk cliffs and at Dowlands there is something of the same kind, but in addition a great fissure, Dowlands Chasm, formed when a landslip occurred at Christmas time, 1839. The chasm itself has changed very little, but chimneys and pillars of rock formed by the slip, which also pushed up the sea-bed, have now weathered away. Other good examples of undercliff occur near Niton and Ventnor in the Isle of Wight, and along parts of the north Yorkshire coast (19).

## DROWNED RIVER VALLEYS

The beauty of our coastline is not confined to cliffs. The numerous inlets and river estuaries around our coasts are lovely features. The estuaries of the Severn and Thames are too large to be appreciated easily as single units, but smaller inlets like Milford Haven, Carrick Roads, The Dart (40), the Teign mouth (38), and the Orwell and Stour estuaries (164) illustrate this point better because, unlike the Thames and Severn, it is possible to obtain a comprehensive view of each of them, and therefore to grasp their form and beauty as a whole.

From the point of view of the geographer they are all drowned river valleys, and their differences depend mainly on the type of country and rock in which they are situated. The soft, sandy rocks of Norfolk and Suffolk intersected by the serene valleys of the Deben, Orwell, and other rivers stand in marked contrast to the more deeply cut valleys of south-western England and Pembrokeshire. Both contrast with the submerged mouths of the Sussex rivers of which the Cuckmere alone remains unspoiled (13, 22, 44, 57).

The origins of these drowned river mouths are largely connected with the events of the great Ice Age. When the ice-caps formed, the water for them must have come from the sea. Calculations based on the estimated extent and volume of these ice-caps suggest that the sea-level may have been 200 ft. or more lower than it is today.

Because the sea-level was lower, much of the continental shelf, that is the belt of land under the sea round the coast of a continent, and covered with water generally less than 100 fathoms deep, must have been exposed; the non-glaciated parts of the rivers could therefore cut their courses to a lower level than they can today. With the subsequent rise of sea-level the lower reaches would have been drowned under the sea, thus producing in main outline the features which many estuaries show at the present time.

The full story is naturally more complex than this, because the change in relative levels of land and sea was not entirely confined to the movement of the sea and because the fluctuation was not a simple up and down one. Nevertheless, on these general lines many of our coastal inlets can be explained.

When the continental shelf stood in part exposed, forests often throve on its slopes.[1] These forests are often seen today near high-water mark, for example on the Norfolk coast

[1] This introduces a difficulty; during the Ice Age, forests could not exist. But it is abundantly clear that forests did exist during times of low sea-level *after* the Ice Age had passed from these latitudes. It has been suggested that the ice was then at a maximum in the southern hemisphere, so that the water was drawn away from the northern area, thus exposing the continental shelf.

and in Sussex. Excavations for docks and other purposes have revealed many forests or peat-beds at lower levels, all clearly indicating a rise by stages of the sea to the present level.

An interesting group of drowned valleys occurs in Norfolk. The Wensum, Waveney, and other rivers in which the Broads lie, were, at the time of the Norman Conquest and perhaps later, open arms of the sea reaching as far inland even as Norwich. Their seaward ends have been blocked by sand- and shingle-spits or bars like those at Yarmouth (159) or at Horsey (before the sea-wall was built). Breydon Water is a natural broad and is the unfilled part of the estuary. Side-valley broads such as Flixton are also natural. But it has now been shown that most, if not all, of the others are man-made. They originated as peat-cuttings (see *Royal Geographical Society, Research Series*, No. 3, 1960).

## RAISED BEACHES, PLATFORMS, WATERFALLS

It has already been stated that the sea once stood at a lower level *relative* to the land than at present; it is also clear, however, that at certain times it stood at relatively higher levels. The proof of this statement lies in the raised beaches (that is, former beaches now above sea-level) and rock platforms on which beaches formerly rested, which are often found round our coasts, more particularly in Scotland. In England and Wales also there are some coastal stretches in which former beaches can be examined, in others there is simply a platform on which the effects of wave erosion are visible and on which the beach originally existed. The south-western coast has examples of both (46). Portland Bill, for instance, has a large spread of pebbles which are the shingle of the old beach.

A noticeable feature in many of the photographs is the plateau-like nature of the cliff-tops. This is exemplified in 81, 85, 89a, b and 123. The origin of this type of landscape has given rise to much discussion and the various authorities have not yet reached agreement. There seems little doubt, however, that the lower surfaces were the work of marine abrasion active at a period when the respective levels of the sea and land were very different from what they are now. Yet it must not necessarily be supposed that the several platforms seen in the illustrations are all connected and part of the same one. Around the coast there are several platforms at different heights above sea-level, and, by way of further complication, because of local movements of land, the same platform may vary considerably in height. In the areas where rock platforms occur, it is sometimes possible to see the terraced nature of the country, the higher platforms which are older and less perfect in form being those farther inland.

In North Devon and Cornwall cliffs cut in plateau and non-plateau areas are distinct, but there is another particular feature which needs comment. Between Lynton and Boscastle (excluding Barnstaple Bay) there is a fine series of coastal waterfalls, of which the full beauty is unfortunately marred because the lower ends of some of these valleys have been altered by man. The main interest lies, however, in the contrast between the almost vertical falls like Litter Water, coming from the flat-topped cliffs and the steep runnels or gutters of the hog's-back cliffs. Often the water has cut an intricate course along softer strata or joints (62, 63).

## ISLANDS

A beautiful feature of our coasts is the number and variety of the large and small islands which surround it. Scotland is particularly rich in the number of its islands, but in England and Wales Holy Island and the Farnes off the Northumberland coast, the Isle of Wight, the Isles of Scilly, Anglesey, and a host of smaller islets and rocks are of great interest (49–52, 59). Many people ask the question 'How are islands formed?' There is no simple and comprehensive answer, and in limited space it is possible only to offer a few hints. As rivers and streams cut their valleys down to sea-level, it may happen that change in the relative levels of land and sea will drown the low-lying land, and the higher parts only may show above the new level of the water as islands. A process of this kind, and the persistent erosion by the sea which naturally accompanies it, means that the islands may have a different appearance now from their original form. The Isle of Wight was cut off in this way; at one time the chalk ridges of Dorset and the island were continuous and the Frome Valley extended along the present Solent and Spithead. Today, of the intervening land Old Harry Rocks (25) and the Needles (21) alone remain. The Menai Strait (108) between Anglesey and Caernarvonshire is the result of the drowning of two river valleys. It is true that the middle reach of the Strait has a somewhat more complicated history involving certain events in the Ice Age, but the general statement is true.

Smaller rocks (86, 87) and islands may be of similar origin. Ramsey Island (90) is merely a part of the Pembrokeshire plateau, and the same is true of Gateholm and Skomer. Others again may be separated largely by marine erosion cutting in along the joints, or by eroding belts of soft rock. Llangwyfan Island illustrates the latter point (109), whereas the islets in 26, 28, 37, 53, 55, 60, 111, and others are for the most part explained by marine erosion.

Sometimes, if the rock structure is favourable, an early stage in the formation of these small islets and pinnacles is the cutting of a natural arch or bridge (39, 85, 133). Incidentally, it is worth remark here that much of the magnificent cliff scenery of Pembrokeshire between Stackpole and Linney Head is not merely the result of marine erosion on a coast made of Carboniferous Limestone, but on a limestone already riddled with caves and hollows through ordinary atmospheric weathering and land water at an earlier stage in its history.

## DUNES AND MARSHLANDS

So far we have been discussing high and cliffed parts of our coast. There are, however, many stretches of dunes and marshlands, shingle spits and sandy forelands, all of which afford many points of great interest. Dunes are formed of sand blown up from the wide beaches exposed at low tide, and much of their interest lies in their close interconnection with plant growth. The common dune-grass *Ammophila arenaria* (marram) thrives best in those places where there is always plenty of sand. The seeds take root on the higher parts of shingle ridges, and once the tuft of grass rises above the ground-level it too becomes a sand-trap. It is thus easy to see how dunes grow: they take on varied forms, depending on their irregular rates of development and also on the effects of wind and sea erosion.

In some places, where there is a wide sandy foreshore, new dune ridges grow up to the seaward of the older dunes. This is very clear in Morfa Harlech (**103**) and many other districts (**158**). But if the beach is narrow or steep and the amount of sand limited, the existing dunes are liable to severe erosion. This applies to much of the Norfolk coast between Happisburgh and Winterton; the great breach made by the sea at Horsey in 1938 was the result of the waves cutting through the narrow dune ridge in a storm. Many more serious breaches were made in the great storm-surge of 1953. Since then many miles of low-lying coast in eastern England have been protected by walls or embankments (**146**). Dune vegetation seldom forms a close cover and the ridges can be readily destroyed; it is always easy for the wind to blow the sand away once a bare patch has been formed round a rabbit-burrow or perhaps as a result of a track worn across the dunes by people walking over them.

There are many fine dune districts around our coasts; apart from those already mentioned there are others between the Ribble and the Mersey (those at Southport have increased very greatly), in many parts of Cardigan Bay, on the coast of South Wales where their encroachments have in the past done much damage, in Barnstaple Bay (Braunton Burrows), and in several parts of Cornwall, notably at St Piran's where a church has been overwhelmed (**24, 74, 100, 107, 120, 151**).

## SHINGLE FORMATIONS

Frequently dunes are an essential part of the landscape in areas of marsh and shingle. The Chesil Beach, Dungeness, and Orford Ness are fine examples of shingle formations, but of these three only the Dungeness district has any dunes. The Chesil is, however, unique (**30**). The beach is continuous from Bridport to Portland, and beyond Abbotsbury runs seaward from the land to enclose the Fleet. The shingle is well and regularly graded. Near Bridport the pebbles are nearly the size of a walnut, whereas as Portland they are about the size of one's fist. Grading is unusual with shingle, and this feature alone makes the Chesil Beach a natural phenomenon. The beach also increases very regularly in height and width towards the south-east, consisting of one main shingle ridge, the seaward side of which is terraced with subsidiary ridges. Each of these subsidiary ridges was probably formed during a particular storm or tide. If, for example, the waves ran far up the beach, they would obliterate all the lower ridges existing before the storm and leave another crest high up on the beach. Only very occasionally are the waves sufficiently high to overtop the beach. On the landward side of the beach there are many hollows or 'cans', caused by the water passing through the beach when the level of the water is higher on the seaward than on the landward side. The inner coast of the Fleet is uncliffed. Whilst it seems certain that the Chesil Beach collects material at both ends, it is not yet possible to give an entirely satisfactory explanation either of the presence of the beach itself or of its peculiarities.

Dungeness and Orford Ness are very different. They both consist of large numbers of individual shingle ridges, each of which was at one time the seaward ridge of the structure (163). It is possible to map the trend of these ridges, and still better to obtain a bird's-eye view of their layout from air photographs. If one ridge cuts across another, the former is clearly newer: similarly the seaward ridges are more recent than those farther inland.

Of the two, Dungeness (6) is the more complicated and juts out in the form of a great triangle enclosing the now reclaimed Romney Marshes. The old cliff-line, formerly the shore, is easily followed, for example, near Winchelsea (7). Orford Ness is the name given to the apex of a great spit of shingle which has deflected the River Ore or Alde for a distance of ten or eleven miles to the south of its original exit. South of Aldeburgh the ridge is narrow, and the course of events if the river were to break through is much debated. It would, however, be difficult for a real breach to occur at Slaughden in present conditions although the beach is narrow. Shingle is generally moving southwards with the beach drift, and rapidly and effectively seals any likely breach (162–3). Even the storm of 1953 failed to make a true breach.

This 'travel' of beach materials, especially shingle and sand, is of the utmost importance, and the most casual observer cannot fail to notice that the beach is often piled much higher against one side of a groyne or breakwater than the other (8).

It is sometimes possible to relate the course of the growth of these shingle formations to the actual history of the locality. It can be shown with reasonable accuracy that in 1162 when Orford Castle was built, at a time of prosperity for the port of Orford, the great spit of shingle had reached only as far south as a place now marked on the one-inch Ordnance map as Stonyditch Point. Thus the spit protected Orford Haven, and was not the menace it became later.

Likewise the relations between the building of Harlech Castle and the growth of Morfa Harlech are traceable, but the evidence of the evolution of the morfa is less clear than that of the growth of Orford Ness. It is, however, a fact that the castle had a watergate, and when it was built presumably a creek of tidal water reached it round the north end of the growing shingle ridge. On account of the changes that have taken place since the spit reached the former island of Llanvihangel y Traethau, and since the sea-wall between that hill and Talsarnau was built in 1812, there is no clear nor certain deduction possible of earlier conditions. Even the information about the former positions of the creeks yielded by air photographs of the area is not absolutely conclusive (103).

There are several other smaller shingle ridges and spits around our coast: that at Slapton in one way resembles the Chesil because it encloses a lagoon. On the North Devon coast there is the famous cobble ridge or Popple at Westward Ho! (64), where the pebbles are particularly large. It has grown to the north and now encloses the golf links. Another small but interesting ridge shuts in the very beautiful Loe pool, near Helston and Porthleven in west Cornwall. On a still smaller scale, but nevertheless of much interest, are the numerous shingle and boulder ridges that now join many small islands and rocks to the bigger islands of the Scilly group (49, 51).

## SALT MARSHES

Within nearly all these features there is often some development of salt marsh, but on those so far mentioned the salt marsh is not the main interest in the area. It is true that there are vast marshes within Dungeness, but they are nearly all reclaimed, as for the most part are those inside Orford Ness. But inside the spits enclosing the Dyfi (Dovey) and Mawddach (100) estuaries in Cardigan Bay, within Hurst Castle Spit (17) and others between it and Pagham harbour on the south coast, along parts of the Essex coast, and especially in north Norfolk, it is the marsh that dominates.

Although there is a great deal of similarity between all salt marshes around our coasts, there is nevertheless much variation in detail and in their general setting. The estuary of the Dyfi, surrounded by mountains and largely blocked by the great shingle and dune spit running north from Borth, stands in marked contrast to the open sand flats with shingle and dune features like Blakeney Point and Scolt Head Island (151–2), which form an outer coast in front of the old cliffs and gently undulating ground landward of the marshes. Further examination of the two areas will soon show that the Cardigan Bay marshes are mainly sandy, whereas those in Norfolk contain a greater proportion of firm mud.

Another striking difference is in the vegetation: those on the west coast are predominantly grassy, whereas although there is much grass on the Norfolk marshes, there is a greater abundance of certain other marsh plants; some, like *Halimione portulacoides* and the sea-lavenders, are often absent from the west-coast marshes (154).

## MARSH FORMATION

Before discussing one or two other details a word about marsh formation in general is necessary. Salt marshes grow in sheltered places, for example, high up estuaries, and behind shingle bars. Originally they were nearly all sand flats exposed at low tide. As the rising tide gradually covered them, it brought with it fine silt and mud suspended in the

8

water. This material may have been carried to and fro many times, but in quiet spots and also at the turn of the tide some was deposited. In this way small mud or silt banks began to form. The original sand 'flat' was not quite level, and seaweed and pieces of flotsam were often scattered over it. Very often mud was deposited round any objects which obstructed the movement of the water, and in this way small mud-banks developed. Many of these were only temporary, but some survived. Hence slowly but surely mud was deposited in various places on the original flats, and these banks would soon become big enough to have considerable effect on the run of the tidal waters, and so increase the amount of the deposits.

Seeds of salt-marsh plants are carried in; some by the flowing tides, others by birds, others by the wind, others again carried unintentionally by man. These may take root, especially in undisturbed areas. Sometimes seaweed on a bank forms a minor protection, or a tiny hollow traps some seed, or perhaps inadvertently man has trodden seeds in the mud and so prevented their being washed away. If they germinate the vegetation soon gets a firm hold. The type of vegetation that grows on a marsh is closely related to its general configuration and to the conditions which obtain upon it, for example, whether the mud is sloppy, firm, or sandy. The type of vegetation also depends on the number of tidal inundations which a marsh receives during the year. Very often the first halophytes (salt-loving plants) are *Zostera* (eel grass or widgeon grass) on wet sloppy mud, and species of *Salicornia* (samphire) on firmer ground. In Norfolk these are followed by the annual *Suaeda*, the sea-aster, and, especially on rather sandy marshes, by the common salt-marsh grass (*Puccinellia maritima*). On many of the west-coast marshes this grass is prolific, and since it is very often grazed, the marshes there generally have a lawn-like appearance (**104**).

At lower levels the marsh is usually covered by all high tides, but where vegetation has taken hold the flow of water is obstructed, thus causing more deposition. Therefore the mud-banks already partly covered with plants grow in height and extent. This process will go on for some time until eventually only the highest tides cover the marsh. Consequently the rate of upward growth will lessen, because the opportunity for mud and silt to settle is more infrequent. At about the time of maximum growth, however, plant life may be prolific, and during the flowering period gives great beauty and colour to the marshes. It is at this stage that the sea-aster (*Aster tripolium*), the sea-pink (*Armeria maritima*), the lavender (*Limonium humile*) and many other plants such as *Triglochin maritimum* and *Spergularia media*, often form a thick carpet divided only by creeks and salt pans (**152**).

At high levels, which are submerged only by high spring tides, there grow the sea-plantain (*Plantago maritima*), wormwood (*Artemisia maritima*), *Glaux*, and *Juncus*. If a river or stream drains into the marsh, the gradual change from maritime to land vegetation is traceable and increases the range of variety.

Before leaving this subject there are two plants which require special mention—*Spartina townsendii*, a marsh grass, and *Halimione portulacoides* or sea purslane. *Spartina townsendii* is a hybrid grass which appeared on the south coast in 1870, and it has spread with remarkable rapidity in soft wet mud. It makes those areas where it grows firmer and less liable to erosion, and because of its rapid increase it is, under control, a valuable ally and is introduced for the purposes of reclamation. But left to itself to grow in a wild condition, it is apt to spread too quickly and even to upset navigation. Parts of Poole Harbour have been transformed in appearance by its growth and it is prolific also in Southampton Water.[1]

*Halimione portulacoides*, or sea purslane, spreads very rapidly over middle-high marshes especially on the east coast (**154**). It likes well-drained areas and begins its growth mainly on creek edges. Thence it spreads over a marsh, virtually excluding all other vegetation. It is not often found on marshes on the west coast.

One very noticeable feature of most salt marshes is the intricate meanderings of creeks through them. It will be remembered how the marsh surface grew upwards, gradually exceeding the level of all but high spring tides. The tidal waters, at first flowing and ebbing as a sheet of water over the bare sand flats, become more and more confined to definite channels as a result of the upward and outward growth of mud-banks now largely plant-covered. In certain local circumstances a creek originally formed in this way is often lengthened by headward erosion (**152, 16, 15, 167, 149, 151**).

Salt-pans develop for much the same reason (**15, 16, 17, 167**). Vegetation does not spread regularly, and often it happens to enclose spaces bare of plants. In the early stages when there is only a sparse covering of plants these bare spots are of no great significance. As the vegetation thickens, the uncovered parts remain below the level of the surrounding marsh on account of the more rapid accumulation of mud round the plants. Because of this evolution the bare patch is one usually undrained. Thus as the tide falls it leaves pools in these places which, with the development of the marsh, become very clearly defined. They usually have no kind of vegetation at all in them, largely because of their stagnancy. If, however, a channel is cut to them, their whole appearance changes in about two years, because they are thus drained and plants can colonize them.

There is therefore much physiographical and ecological interest in marshes, and there is also great beauty. The varying sober colours of the plant-covering at any time of the year, the glorious spreads of colour in the flowering seasons, the ever-changing cycle of the tides, and the limitless views and expanses of sky, give to marshes an unrivalled attraction.

Sand-dunes, shingle-spreads, and marshes have been discussed separately, but they are most often found together. At Blakeney Point and Scolt Head Island (**151–2**), for instance, shingle-ridges originally formed by wave-action on the sand-flats have lengthened by beach-drifting, and have pushed out a number of lateral shingle-ridges, each of which was formerly the end of the main ridge. Between these laterals, and also between the main

---

[1] The grass which is referred to in this paragraph is not strictly *Spartina townsendii*, but a derivative of that plant. I think it has not yet been given a specific name.

ridge and the old land, marshes have developed, while dunes sometimes formed on the shingle-ridges. Conditions in such places are always changing. The waves are constantly working on the beach and tending to lengthen the main ridge, even though storms may often cut off part of it. But sooner or later the ridge grows again. The main beach itself is slowly moved landward by waves washing over it in storms. This is often clearly demonstrated by the occurrence of marsh mud on the foreshore: the dunes and shingle have overstepped the marsh which now reappears on their seaward side. Changes, too, in the dunes are evident; new ones grow around marram grass, whereas older ones are blown away. The marshes meanwhile are passing through the normal stages of development which have been outlined. Accretion and erosion are concurrent processes in a marsh area as a *whole*, but either one may for a time predominate in a particular *part* of it (see also 167).

## THE INFLUENCE OF GEOLOGICAL STRUCTURE

The nature of any stretch of coast is closely related to the structure and form of the land itself. In general, the north and west of England are formed of older and harder rocks than the south and east. This is not, however, always clear from a glance at a map. Thus the ancient rocks of the Lake District are bordered by a fringe of newer rocks and often by boulder-clay, so that Cumberland has in reality a coast of 'soft' rocks (118, 119, 124). Lancashire, including Morecambe Bay, is a low-lying area of soft and newer rocks, together with much boulder-clay. Hence, we find wide sand-flats (117) and extensive dune areas.

Much of the coast of Wales is formed of hard rock, even if a boulder-clay fringe like that near Rhyl and Prestatyn, or that plastering the cliffs just south of Harlech is evident. Where the coast is rock-bound, the detail depends very much on the actual structure of the rocks. This is particularly clear in Pembrokeshire and the Gower peninsula. It is, on the whole, true to say that all the headlands between Fishguard and St David's are formed of igneous rocks, while the bays are cut in the softer sedimentaries (Fig. 11). South of St Bride's Bay the rocks, like those in the Gower peninsula, are folded along roughly east and west lines so that inlets like Milford Haven and that between Gower and Llanelly lie along the strike of the rocks. The detail of a coastline is also closely bound up with rock structure. Freshwater West and Freshwater East bays in Pembrokeshire are at either end of the outcrops of Old Red Sandstone, and correspond in position to Rhossili and Oxwich bays in the Gower peninsula (Fig. 9).

A similar east and west strike of the rocks determines much of the coastal scenery of Dorset and the Isle of Wight (Figs. 3 and 4). The chalk ridge running from Culver Cliff to the Needles (21) finds its continuation in Old Harry Point (25) and so via Corfe Castle to Worbarrow Bay, Lulworth Cove, and White Nothe (29). The soft rocks hollowed out in Swanage Bay are comparable with the east and west widenings of Lulworth Cove (28) and Worbarrow and Mupe bays (26). Similar features appear on the east and west sides of the Isle of Wight. Both in Dorset and Pembrokeshire many of the smaller inlets are arranged at right angles to the east–west trend of the rocks. In some localities they may be dependent upon faulting, for example, Swanlake Bay and Manorbier Bay (**83**), or upon the action of sea erosion on joints as in the rocks at Stair Hole.

In most of the great peninsula of Devon and Cornwall the rocks are arranged in folds of which the general direction is east–west (Fig. 6). But the folding is very ancient, and large igneous masses like that of Land's End often break into this arrangement. Although in this instance it is only approximately true that structure has any close relation to coastal form, nevertheless igneous and hard sedimentary rocks frequently make headlands, whereas the softer rocks have been cut back to form bays.

Vertical movements have also been important and in the east and south-east of England this is at once apparent in the beautiful estuaries of Suffolk and Essex and those now dammed by shingle and in many cases spoilt by man in Sussex.

But these areas are formed of softer rocks and erosion has destroyed nearly all traces of former raised beaches. For this reason it can be said that the cliffs and shingle, and the dunes and marshes, are largely the product of modern conditions. Recent geological formations indeed obtain in East Anglia as a whole. Between the chalk of Flamborough Head (**142**) and Hunstanton (**150**) great changes have taken place since the period before the Ice Age. At that time the York and Lincoln Wolds formed the coast, and the sea ran up into the Humber marshes and over all Holderness as well as over the Fenland. Holderness 'bay' is now filled with boulder-clay and the sea is eroding the cliffs. The Fenland was partly filled with boulder-clay, partly clogged by peat growth and the formation of the fen clay, and finally reclaimed by man. The Lincolnshire coast (**146–8**) is largely a dune one covering the boulder-clay filling.

To the north of Flamborough there have been similar changes. Filey Bay is plugged by boulder-clay (**141**) and the Derwent river now flows in a direction opposite to that of its former course. North of Filey the Yorkshire coast is made up of Jurassic rocks of greatly varying resistance to the sea and the weather. The bays are often filled with boulder-clay, but the present form of the coastline results largely from the effects of constant action on the various rocks by the sea (**135–9**).

North of the Tees, where solid rocks reach the coast, the resistance to erosion has been considerable, producing thereby the fine cliffs in the Magnesian Limestone of Durham (**132–3**) and those in the Coal Measure sandstones of part of Northumberland. Between Cullernose Point and Dunstanburgh Castle the igneous rock of the Whin Sill (**129, 130**) is the main feature. In Northumberland, however, one of the most characteristic features is the number of wide bays (**128**) backed by sand dunes which lie between outcrops of harder rocks. Some at least of these bays coincide with faults, and small streams find their way into the sea through the dunes.

## PLANNING TO PRESERVE

In these few pages and photographs some of the outstanding features of our really beautiful coast are described. This is no place to discuss future planning of the use of our coasts, but it is perhaps allowable in the final paragraphs to call attention to the importance of preserving a most valuable national heritage.

Toward the end of the war I made a complete survey of the coast from the point of view of amenity and use. Detailed reports were compiled, together with a map, expressing my own assessment, based on the simplest principles, of the beauty, use, and misuse of our coast.

Even those who have paid only rare visits to the seaside will realize that many miles of the coast are already spoiled by ill-planned and ill-sited bungalows and villas. Other parts are ruined by unsightly mining and manufacturing development or by past and present quarrying. At the present time nuclear stations are an additional menace! In fact it would have been possible to replace all the illustrations in this book with others showing disfigured parts of the coast.

There is no need to spoil the coast when it is made more accessible to the visitor. But foresight and careful planning are required, especially in view of the greater popularity which the future constantly brings to the coast. Camps, both big and little, new houses and bungalows can be built, provided they are properly sited. Everyone is familiar with the defects of ribbon development along our many main roads: surely it is an even greater tragedy that our only coast should be lined with bungalows and houses. More and more people wish to enjoy the *natural* beauties of the coast. They will be able to do so with greater satisfaction to themselves if systematic planning takes place, for without planning the spoiling of the coast will continue. But planning must include all tourist amenities, including car-parks. Every year the problem becomes more difficult, and it is largely because we have failed to take an overall view.

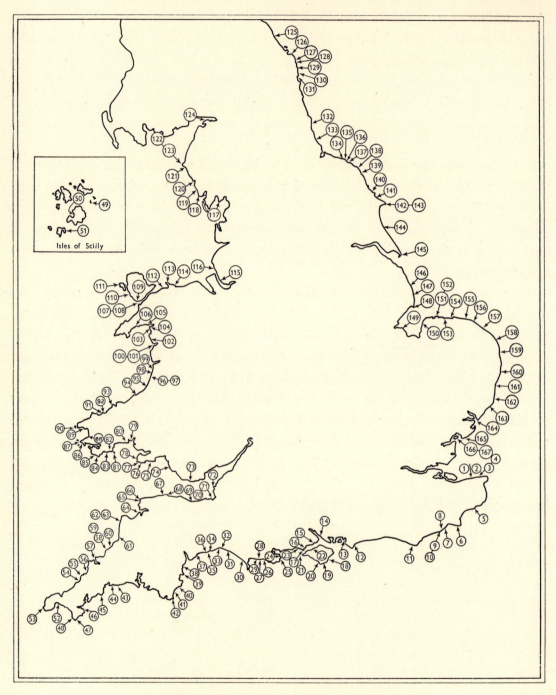

Map showing the site of each photograph.

# NOTES ON THE PLATES

1 North Kent coast at Reculver. The twin towers of the great monastic church of St Mary stand within a Roman coastal fort of the late third century, the northern part of which has been destroyed by erosion.

2 Reculver. Erosion is severe and the church towers are preserved by strong sea defences.

3 Birchington Bay, showing erosion of the chalk cliffs and the gently sloping foreshore exposed at low tide.

4 Birchington, Chalk cliffs. (Photo: British Council.)

5 Steep cliffs in Chalk, and landslips, at The Warren, Folkestone.

6 Dungeness, looking north-east, showing the shingle ridges picked out by growth of vegetation.

7 Winchelsea. Former sea-cliff of Ashdown Sand, and the Strand Gate. Alluvial meadow and marsh in foreground. (Photo: Geological Survey.)

8 Pett Level, and the abandoned sea-cliff near the Coast Guards' Station (now much altered along the new sea-wall). (Photo: Aerofilms Ltd.)

9 Cliffs south-east of Fairlight Glen, Hastings. The cliffs are formed of Ashdown Sand and Fairlight Clay, partly slipped on the light slope in the middle of the photograph. (Photo: Geological Survey.)

10 Cliffs below Ecclesbourne Glen, Hastings. The ravine is cut in Ashdown Sand. (Photo: Geological Survey.)

11 The Chalk cliffs near Birling Gap, looking toward the Seven Sisters. (Photo: British Council.)

12 Sandspits impounding the shallow estuary known as Pagham Harbour.

13 Chichester Channel, looking south-south west.

14 The Hamble estuary at low tide.

15 *Spartina* salt marsh in the estuary of the Beaulieu river.

16 Salt marshes at Keyhaven, looking south-west.

17 Hurst Castle Spit and salt marshes at Keyhaven.

18 Luccombe Beach: in the distance Culver Cliff, Isle of Wight. (Photo: British Council.)

19 Coastal landslips at Rocken End, Isle of Wight. Chalk overlying Lower Cretaceous beds, dipping gently seawards.

20 Whale Chine, Isle of Wight. The view is taken looking towards the sea. The chine is cut in recent alluvium and gravel resting on Lower Greensand. (Photo: Geological Survey.)

21 The Needles: stacks in steeply dipping Chalk of the Isle of Wight monocline.

22 The estuary of Newtown River, Isle of Wight, looking south. The decayed medieval borough of Newtown lies beside the left-hand branch of the estuary.

23 Barton Sands: the eastern part of Christchurch Bay. (Photo: Eric Kay.)

24 Studland Heath. Sandy heath and dune country on Tertiary beds, on the south side of Poole harbour, looking south.

25 Stacks and vertical cliffs in Chalk, Foreland Point.

26 The Dorset coast, eastwards from Lulworth. (Photo: Aerofilms Ltd.)

27 The Fossil Forest, east of Lulworth Cove. (Photo: Eric Kay)

28 Lulworth Cove and the coast of Dorset to the west. (Photo: Aerofilms Ltd.)

15

29  The Dorset coast, Chalk cliffs, west from Durdle Door. (Photo: Eric Kay.)

30  The Chesil Beach. (Photo: *The Times*.)

31  Cliff of Bridport Sands and Inferior Oolite, Bridport. (Photo: Eric Kay.)

32  Looking towards Golden Cap, Lyme Bay. (Photo: Eric Kay.)

33  Coastal landslips, Black Ven (Greensand and Gault sliding forward over Lower Lias).

34  The mouth of the Axe, slightly deflected by a shingle spit, Seaton. (Photo: Eric Kay.)

35  Seaton Bay, looking south-west to Chalk cliffs at Beer Head.

36  Under Hooken, south of Beer, Devon. Landslip at Hooken Cliffs, foundered masses of Chalk. (Photo: Geological Survey.)

37  Ladram Bay, near Sidmouth, showing stacks. (Photo: Aero Pictorial Ltd.)

38  Sand bar almost blocking the Teign estuary, at Teignmouth: view towards the south-west.

39  Corbyn (Corbon's) Head, Torquay. Caves and a natural arch cut in Permian rocks. (Photo: Geological Survey.)

40  Drowned valley: looking north up the Dart estuary. Cliffs composed of Devonian slates and mudstones.

41  Bee Sands. (Photo: British Council.)

42  Start Point. (Photo: Aero Pictorial Ltd.)

43  Lantivet Bay, South Cornwall. (Photo: British Council.)

44  The drowned estuary of the River Fowey: view inland over Polruan.

45  The Dodman. (Photo: British Council.)

46  Nelly's Cove, Porthallow. Obliquely stratified sandy and gravelly Head on raised (*Patella*) beach deposits resting on rock platform eroded in black slates (Killas) and cherts of Veryan Series. (Photo: Geological Survey.)

47  Looking toward Basse Point from Kennack Bay; the Lizard platform. Hot Point (in the distance) is formed of hornblende schist; Enys Head (middle distance) is made of serpentine, which also forms the reefs in the bay. A storm beach in the foreground with blown sand behind. (Photo: Geological Survey.)

48  Kynance Cove is cut in serpentine rocks, and its detail is partly due to erosion along faults. (Photo: British Council.)

49  Isles of Scilly: a general view of the Eastern Isles from Great Arthur. Note the tied island. (Photo: James Gibson.)

50  Isles of Scilly: the north coast of St Mary's Island and the Interior Sea. (Photo: James Gibson.)

51  Several of the islands in the Scillies are joined by shingle-bars formed by wave-action. The photograph shows that joining St Agnes (left) to Gough. (Photo: J. A. Lofthouse.)

52  St Michael's Mount. View looking across to the mainland, at low tide.

53  Land's End. The castellated character of the granite is emphatic. (Photo: British Council.)

54  Perran Beach.

55  View north-east over Kelsey Head to West Newquay. Indented coastline in Devonian slates.

56  Watergate Bay looking south-south-west to Trevelgue Head. Cliffs in Devonian slates. Note the Iron-Age cliff-top fort in the right foreground.

57  The Camel estuary. View from the south-east past Padstow (the harbour is on the right). Country rock here composed of Devonian slates.

58  Tintagel Head. The photograph shows the scattered buildings of the famous Dark Age settlement on this headland which is all but separated from the mainland.

59  Rocky Valley Mouth, Tintagel. The gorge is at the mouth of a hanging valley cut in Upper Devonian Phyllites. (Photo: Geological Survey.)

60  Boscastle Harbour. The sea has cut back into the land along joints. (Photo: Geological Survey.)

61  Active erosion in progress on the north Cornish coast, south of Bude, here composed of steeply dipping Culm Measures.

62  View-point, Speke's Mill Mouth (Milford Water), Hartland Quay. The shore-reefs are

composed of the basset edges of Culm Measures sandstone. The cliffs in the foreground consist of a limb of an anticline that dips towards the observer; the other limb is seen near the pinnacle—St Catherine's Tor—where a second anticline begins. Part of the valley behind the Tor is visible. (Photo: Geological Survey.)

**63** Waterfall, Milford Water, Speke's Mill Mouth. The fall is down the limb of a sharp syncline in Culm sandstones, and at the foot the stream is flowing along the strike. (Photo: Geological Survey.)

**64** The pebble ridge, Westward Ho! (Photo: Geological Survey.)

**65** Woolacombe Sands, Morte Bay. View towards the south towards the headland that forms Baggy Point (Devonian and Old Red Sandstone).

**66** Bull Point really forms a right-angled turn in the coast, and is hardly a true promontory. It shows jagged reefs of Morte slates. (Photo: British Council.)

**67** Valley of the Rocks, Lynton. The now streamless valley runs almost parallel with the coast where it is bounded by a rugged wall of hard weathered Lynton shales and grits. Castle Rock is at the end of the road. (Photo: Geological Survey.)

**68** North Devon coast at Foreland Point, here composed of steeply dipping Devonian rocks.

**69** Blue Anchor Point, Somerset. The cliffs are formed of Trias, capped by Rhaetic and Lias. The rocks have been folded into a shallow anticline. At the axis of the anticline the vertical headland—Blue Anchor Point—is produced by undercutting and collapse of horizontal beds, the vertical profile being mainly due to marine erosion.

**70** Rock ledges (Lower Lias), picked out by seaweed at low tide. Coast of Bridgwater Bay.

**71** Mud-flats at low tide. Steart Island: view towards the south-east.

**72** Brean Down, an outlier of Carboniferous Limestone, looking east over the south end of Weston Bay.

**73** Breaksea Point, looking west at low tide. Clays and limestone of the Lower Lias form the coast.

**74** Sand dunes, Kenfig Burrows, as in 1948; they now look very different because of the erection of a steel works.

**75** Pwll-du Bay, Gower peninsula. Storm beaches are well shown. (Photo: Geological Survey.)

**76** View towards the west along the south coast of the Gower peninsula to Worm's Head. Cliffs cut in well-bedded Carboniferous Limestone.

**77** The coast south-east of Foxhole Slade, Port-Eynon, Gower peninsula. The plateau and cliffs are carved in highly inclined beds of Carboniferous Limestone. (Photo: Geological Survey.)

**78** Rhossili Bay, Gower peninsula. Terrace of glacial drift backed by Old Red Sandstone hills. Sand dunes in the distance. (Photo: Geological Survey.)

**79** The estuary of the River Taf: a view towards the north-west over Wharley Point.

**80** Coastal foreland of recent geological date, fronting the old line of sea cliffs cut in Old Red Sandstone.

**81** View towards the north-north-west over Caldy Island, showing a well-developed plain of marine erosion.

**82** Lydstep Point: cliffs in vertically bedded sandstone of Millstone Grit age.

**83** Looking west across Manorbier Bay: the coastline is eroded out of Old Red Sandstone beds that lie nearly vertical.

**84** Huntsman's Leap, Bosherston, Pembrokeshire. A vertical-fault gash in a Carboniferous Limestone cliff. The gash has been eroded by the sea along a fault. (Photo: Geological Survey.)

**85** The Green Bridge of Wales, Pembrokeshire. The arch and caves are cut in the Carboniferous Limestone, and the plateau is beautifully shown. Saddle Head is in the distance. (Photo: Geological Survey.)

**86** Gateholm Island at the west end of Marloes Bay.

**87** View east from Skomer Island (Ordovician volcanic rocks) over Midland Isle to Wooltack Point.

**88** The Sleek Stone, Broad Haven, Pembrokeshire. A monocline in Coal Measures Sandstone. (Photo: Geological Survey.)

**89** Coast scenery at Solva, Pembrokeshire. The drowned valley is cut through Middle Cambrian rocks. The plateau nature of the interior is well shown. (Photo: Geological Survey.)

**90** Looking south-west across St David's Head to Ramsey Island. Erosional outliers of Carn Ffald and Carn Llidi in foreground.

**91** Looking over Pwll Gwaelod to Newport Bay. Cliffs in Llandovery mudstones.

**92** The coast at Cwm yr Eglwys (Newport Bay) Pembrokeshire. (Photo: British Council.)

**93** Folding in Ordovician mudstones. Coastal cliffs south of Cemmaes Head.

**94** The coast of Cardigan Bay looking toward New Quay from above Aberayron. (Photo: British Council.)

**95** Foreshore, fronting narrow coastal plain, near Llanrhystyd.

**96** The mouths of the Rheidol and Ystwyth at Aberystwyth. (Photo: Aerofilms Ltd.)

**97** Cormorant Rock, near Aberystwyth. A sea stack on the shore platform cut in Aberystwyth Grits, which hereabouts consist of regular alternations of evenly-bedded grits and shales. (Photo: Geological Survey.)

**98** Clarach Bay, near Aberystwyth. This is one of the smaller drowned valleys running to Cardigan Bay. The stream, a misfit, is deflected by a shingle-bar. (Photo: British Council and Batsford.)

**99** Coastal cliffs, $2\frac{1}{2}$ miles north of Aberystwyth, with well-developed bedding planes in Silurian mudstones.

**100** Sand-spit, Ro Wen, enclosing the Mawddach estuary. Looking south-west over Barmouth Bay.

**101** The Mawddach Estuary at low tide.

**102** Looking north over Barmouth along the west fringe of the mountains of the Harlech Dome. Sand-dunes of Morfa Dyffryn and Morfa Harlech in the distance.

**103** Morfa Harlech and the Snowdonian mountains. (Photo: J. A. Steers.)

**104** A creek on Talsarnau marshes; compare Plate 154. (Photo: J. A. Steers.)

**105** Criccieth Castle.

**106** Hooked sand-spit at Pwllheli: view south-west towards Trwyn Llanbedrog.

**107** Looking north over the sand-spit at Aber Menai point towards Newborough Warren.

**108** Panorama of Menai Strait. View to the north-east over Port Dinorwic.

**109** Llangwyfan Church, Anglesey, on an islet of boulder-clay. (Photo: Geological Survey.)

**110** Tre-arddur Bay on the west coast of Anglesey, here composed of rocks of the Mona Complex. Short sandy beaches, at the heads of small bays.

**111** South Stack, Holyhead. The highest cliffs in the two islands are hereabouts. (Photo: *The Times*.)

**112** View east over the low-lying coastal strip to Allt Wen and Conway Mountain, composed of Ordovician volcanic rocks.

**113** Cliffs in almost horizontally-bedded Carboniferous Limestone.

**114** Aerial view of Colwyn Bay showing the Little Orme and the Great Orme. Tan Penmaen Head in the middle distance. (Photo: Aerofilms Ltd.)

**115** Dingle Point, south of Liverpool. The low cliffs (facing the Mersey) are of Upper Mottled (Bunter) Sandstone. In the foreground is a marine-eroded bench. (Photo: Geological Survey.)

**116** Blundellsands foreshore: the River Alt formerly changed its course frequently and by flowing along the coast led to serious erosion. The photo shows the wide expanse of sand characteristic of this shore. (Photo: Geological Survey.)

**117** View from Park Head. The topography is composed of Carboniferous Limestone (middle distance) and a fault escarpment of Silurian rocks in the background. Part of the Leven estuary is shown. (Photo: Geological Survey.)

**118** The foreshore between Gutterby Spa and Summer Hill, two miles north of Silecroft. The cliffs are of glacial material, mainly sand and gravel above with loamy sand and laminated clay below. Probably they rest on boulder-clay. Height of cliffs 100–150 ft. (Photo: Geological Survey.)

**119** Annaside Banks; the photograph shows details of the erosion of cliffs in boulder-clay.

**120** The Irt, Mite, and Esk form a single estuary, deflected in part to the south by the sand-spit of Drigg Point, and in part to the north by the sands at Eskmeals. The details of its evolution are obscure. (Photo: Geological Survey.)

**121** The old course of the Rottington Beck, St Bees. The beck now turns westward beneath the distant scar, and reaches the sea through a storm beach. The small, isolated hill is formed of glacial sands and gravels resting on boulder-clay. The cliffs in the background are of St Bees sandstone capped by mid-glacial sands and gravels. (Photo: Geological Survey.)

**122** St Bees Head. The large boulders in the foreground form one of the scars of this coast. (Photo: British Council.)

**123** Cliffs in well-bedded St Bees sandstone of Triassic age.

**124** The foreshore between Bowness and Solway Bridge, showing marine warp with a narrow strip of raised beach and warp (by the road). Boulder-clay occurs in the background. (Photo: Geological Survey.)

**125** Foreshore, Green's Haven. Syncline in strata of Middle Limestone Group (Lower Carboniferous). (Photo: Geological Survey.)

**126** View south-east over Bamburgh Castle, showing the sand dunes that surround the rocky knoll on which the Castle stands.

**127** Foreshore north of Beadnell Harbour: Whin dyke in foreground and white sandstone underlying Sandbanks limestone behind. (Photo: Geological Survey.)

**128** Beadnell Bay: dunes. (Photo: Geological Survey.)

**129** The scarp and dip slopes of the Whin Sill are well seen. In the distance the small harbour of Craster: in the foreground Dunstanburgh Castle, important in the history of Northumberland in Dark-Age and Norman times.

**130** Cullernose Point: contorted limestone and columnar Whin. (Photo: Geological Survey.)

**131** Panorama towards the south over Alnmouth to Warkworth Harbour, at the far end of the bay. Alnmouth was planned as a new borough and port in the twelfth century, but lost its harbour owing to a change in the river course.

**132** Holey Rock, Roker (as it was), upper thin-bedded (Permian) limestones with caves. (Photo: Geological Survey.)

**133** Black Hall Rocks. Bedded (Permian) Dolomites on Brecciated Reef Limestones. Same succession in the stack. (Photo: Geological Survey.)

**134** Boulby: the cliffs show Lower Lias (at base), Middle Lias and boulder-clay. The farthest point is Staithes Nab. (Photo: Geological Survey.)

**135** Staithes and the fine Lias cliffs north thereof. (Photo: British Council.)

**136** Port Mulgrave. The Jet Rock forms the foreshore, the cliff is of Upper Lias, Alum Shales, Dogger and Estuarine Beds with the Eller Beck Bed near the top of the highest part. (Photo: Geological Survey.)

**137** Runswick Bay. (Photo: Will. F. Taylor and Batsford.)

**138** Saltwick Bay. The cliff is Alum Shale (Upper Lias), Dogger Sandstone, and Estuarine Beds. The outlying stacks forming Black Nab are Alum Shale only. Beyond the Nab the waves are breaking on the Jet Rock, exposed at lowest tides. The platform at the foot of the cliffs opposite the Nab is the floor of an old alum works. (Photo: Geological Survey.)

**139** Robin Hood's Bay, enclosed north and south by steep and lofty cliffs, and backed by cliffs of boulder-clay which give place to the fine sweeps of Fylingdales Moor. Within the bay ledges of Lias are shown. (Photo: British Council.)

**140** Cliffs at Yons Nab. The direction of the view is inland, towards the south-west. Cliffs in gently dipping Cornbrash, Oxford Clay and Corallian Beds.

**141** Filey Bay.

**142** Flamborough Head. (Photo: Aerofilms Ltd.)

**143** Flamborough Head is formed of Chalk covered with boulder-clay. A blow-hole is shown. (Photo: Geological Survey.)

**144** The coast of Holderness. View to the north-west, past Grimston. Active erosion of cliffs in boulder-clay.

**145** Spurn Head: view towards the north.

**146** The eroded foreshore near Sutton, Lincolnshire, after the great storm of 1953. (J.A.S.)

**147** Dunes and beach cusps, Lincolnshire coast. (Photo: H. H. Swinnerton.)

**148** Sand ridges, Gibraltar Point. View towards the north, to Skegness.

**149** Mud-flats at low tide, east of the Welland outfall: Holbeach St Matthew.

**150** Hunstanton Cliffs: White Chalk ('Sponge' Bed at base), Red Chalk, and Carstone (Lower Greensand). (Photo: Geological Survey.)

**151** Scolt Head Island; panorama from the east.

**152** Shingle laterals and salt marsh, Scolt Head Island.

**153** Wells-next-the-Sea. Reclaimed marshes and Holkham woods on left: natural marshes on right.

**154** Morston marshes: Blakeney Church in background. (Photo: Geological Survey.)

**155** The cliffs of glacial materials west of Sheringham. Slips are common and the sea gradually removes the fallen material. (Photo: British Council.)

**156** Cliffs in boulder-clay on the Norfolk coast. View over Overstrand from the north-west.

**157** Bacton: a medieval wall exposed by erosion of the cliffs. (Photo: Hallam Ashley.)

**158** Winterton sand-dunes from the north-north-west.

**159** Great Yarmouth: the mouth of the Yare is deflected by the spit.

**160** Benacre Broad, dammed by sea beach.

**161** Minsmere Beach; new planting following 1953 flood.

**162** River Alde at Slaughden. View south along the coast to Orford Ness.

**163** Orford Haven. View from the south-west towards Orford Ness.

**164** The Orwell Estuary, near Freston Hall. London Clay on the foreshore. (Photo: Geological Survey.)

**165** Sand spit and salt marshes, Colne Point.

**166** West Mersea.

**167** Sand ridge and mud-flats at low tide, Ray Creek.

# The Plates

1 Reculver

2 Reculver cliffs

3　Birchington Bay

4　Birchington

**5** The Warren, Folkestone

**6** Dungeness

7 Winchelsea, former cliff

8 Pett Level

**9** Fairlight Glen

**10** Ecclesbourne Glen

11 Birling Gap

12 Pagham, from Selsey

13 Chichester channel

14 The Hamble estuary

15 Beaulieu estuary

16 Keyhaven Marshes

**17**   Hurst Castle spit

**18**   Luccombe beach, I.O.W.

**19** Rocken End, I.O.W.

**20** Whale Chine, I.O.W.

21 The Needles, I.O.W.

22 Newtown River, I.O.W.

23  Barton sands

24  Studland Heath

**25** Old Harry Point

**26** Worbarrow Bay

27 The Fossil Forest, east of Lulworth Cove

28 Lulworth Cove

**29** West from Durdle Door

**30** Chesil Beach

31   The cliffs at West Bay

32   Towards Golden Cap

**33** Black Ven

**34** The entrance to the Axe, Seaton

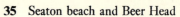

36 Under Hooken

37  Ladram Bay

38  Teignmouth

**39** Corbyn Head

**40** The mouth of the Dart

**41** Bee Sands

**42** Start Point

**43** Lantivet Bay

**44** Fowey River

**45** The Dodman

**46** Nelly's Cove, Porthallow

**47** Kennack

**48** Kynance Cove

**49** Isles of Scilly

**50** Isles of Scilly

**51** Isles of Scilly

**52** St Michael's Mount

53 Land's End

54 Perran beach

**55** From Kelsey Head

**56** Watergate Bay

57 The Camel estuary

58 Tintagel

59   Rocky Valley, Tintagel

60   Boscastle harbour

**61** South of Bude

**62** Speke's Mill Mouth

63 Waterfall, Speke's Mill Mouth

**64** Westward Ho!

**65** Woolacombe Bay

**66**  Bull Point, Morthoe

**67**  The Valley of the Rocks, Lynton

**68**  Foreland Point

**69**  Blue Anchor Point

**71** Steart Island

72  Brean Down

73  Breaksea Point

**75** Pwll-du Bay

**78** Rhossili Bay

**79** Llanstephan peninsula and the Taf estuary

**80** Laugharne

**81** Caldy Island

**82** Lydstep

**83** Manorbier

**84** Huntsman's Leap

**85** The Green Bridge
of Wales

**86** Gateholm Island

**87** Deer Park peninsula, Midland Island, Skomer

**88** The Sleek Stone

89 *a*  Solva

90  Ramsey Island

89 *b*   Solva

91   Dinas 'Island'

**92** Cwm yr Eglwys

**93** Cemmaes Head

**94** Coast, Aberayron to New Quay

**95** Coast near Llanrhystyd and Afon Wyre

96 Aberystwyth

97 Cormorant Rock, Aberystwyth

**98**   Clarach Bay

**99**   Between Borth and Aberystwyth

**101** The Mawddach estuary

**102** From the Mawddach
to Harlech

**103** Morfa Harlech

**104**  Talsarnau Marshes

**105**  Castle Headland, Criccieth

**106** Pwllheli

**107** Aber Menai Point

**108** The Menai Straits

**109** Llangwyfan church

**110** Tre-Arddur Bay

**111** South Stack, Holyhead

**113** The Great Orme

**114** Colwyn Bay

**115** Dingle Point

**116**  Blundellsands

**117**  From Park Head

**118** Foreshore near Silecroft

**119** Near Annaside

**120**  The Irt, Mite and Esk estuary

**121**  Rottington Beck, old course

**122** St Bees Head

**123** St Bees Head

**124** Foreshore near Bowness

**125** Green's Haven

**126** Bamburgh Castle

**127** Near Beadnell harbour

**128** Beadnell Bay

**129** Dunstanburgh Castle

**130** Cullernose Point

**131** Alnmouth

**132** Holey Rock, Roker

**133** Black Hall Rocks

**134** Boulby

**135** Staithes

**136** Port Mulgrave

**137** Runswick Bay

**138** Saltwick Bay

**139** Robin Hood's Bay

**140**  Yon's Nab

**141**  Filey Bay

**142** Flamborough Head

**143** Flamborough Head

**146** Near Sutton, after the 1953 floods

**147** Chapel Point

**148**  Gibraltar Point

**149**
Salt marshes of the Wash

**150** Hunstanton cliffs

**151** Scolt Head Island

**152**   Scolt Head Island

**153**   Wells-next-the-Sea

**154** Blakeney Marshes

**155** West Cliff, Sheringham

**157** Bacton: an exposed medieval well

158 Winterton Ness

159 Great Yarmouth

**160** Benacre Broad

**161** Minsmere beach

**164** The Orwell estuary

**165** Colne Point

**166** Mersea Island

**167** Mersea Island

# REGIONAL COMMENTARY

## RECULVER TO BRIGHTON
### (PLATES I–II)

There are many features of great interest and significance along this stretch of coast. The key to the arrangement of the strata forming the several lines of cliff is given by the folding to which the whole district was subjected in Tertiary times. The Weald is a complicated arch. The Chalk cliffs between Dover and Deal are part of the northern limb of the arch, those at Beachy Head part of the southern limb. The Isle of Thanet represents a minor upfold, between which and Deal is the low ground followed in part by the Wantsum Channel which formerly made Thanet a true island.

Between Dover and Eastbourne the older beds reach the coast—it would be more correct to say reached the coast, because much of the modern coast is formed of the great shingle spreads, enclosing valuable marshland, of Dungeness and Langney (Langley) Point or the Crumbles. The older beds are very varied in composition. The Gault, a soft, sticky clay, is easily eroded and lends itself to landslips. The Warren at Folkestone, where the Chalk slips on the Gault, is the best example. The precise cause of the slips is not certain, but modern views suggest that they are produced by the erosion of the 'toe' of the Gault by the sea, and that the slip takes place along an axis that is roughly horizontal. The cliffs near Hastings are made of beds of that name, but are subdivided into small groups. The Ashdown sand and sandstones are responsible for Castle Hill; at Fairlight there are alternating clays and sandstones much overgrown and locally slipped. The Ashdown sands reappear at Cliff End.

The Chalk, however, is the most spectacular rock in south-eastern England, and makes fine cliffs in Thanet, between Deal and Dover, and from Beachy Head to Newhaven and Brighton. We shall meet the Chalk on other parts of the coast and it may help to give a short general account here before describing particular cliffs. Chalk is a soft limestone, but, in the landscape, it behaves very differently from other limestones. It is usually well bedded, and well jointed. Locally two sets of joints are at right angles to one another, and so if the bedding is fairly horizontal, stacks and chimneys are formed on the coast. Inland the Chalk often forms ridges, like the North and South Downs. These, and the Chilterns and the Lincolnshire Wolds, are escarpments—the Chalk having been tilted up and much worn away by erosion. The contact between the inner and outer facing slopes of the Chalk enclosing the Weald illustrates the difference between the dip and the scarp slopes. Chalk country nearly always carries a turf of short springy grass, and the hills are rounded and bare of trees except where spreads of clay or gravel overlie them. In those places beech groves are characteristic.

The Chalk usually makes steep or vertical cliffs. To some degree the slope will depend upon the inclination, or dip, of the beds, but even so there seems to be a fairly close balance between the effects of marine and subaerial erosion so that the combined effect of both produces a nearly vertical cliff. The skyline of the cliff will depend much more upon other factors. The relatively flat Isle of Thanet gives cliffs with a level skyline, but the jointing in the Chalk adds interesting detail in the form of stacks and gullies. The former extent of the cliffs is shown by the erosion platform which extends in front of them. Between Deal and Dover the cliffs pass through the whole northern limb of the Chalk; on the northern side they show the slopes of the dip, on the southern side of the scarp. Their height largely depends on those factors.

109

At Beachy Head the cliffs are vertical and reach a height of 500 feet. The south-westerly-facing side bears the brunt of the attack from the sea. Between the headland and Eastbourne the structure is complicated, and the Upper Green-sand and Gault both play a part. Moreover, there are also several faults. To the west of the headland the cliff profile is well known and forms the Seven Sisters. This line of cliffs is parallel to the strike of the Chalk which here was formerly drained by valleys on its dip slope when the Chalk land extended farther seawards. As the sea cut into the Chalk, the valleys ceased to function, but their upper parts show as depressions in the cliff profile.

The Chalk forms the cliffs almost as far as Brighton. Seaford Head is conspicuous, and is probably the last remnant of a fold. Nearer Brighton the cliff top is fairly level, and the Chalk has suffered a good deal of erosion. Today there are extensive protection works, especially near Brighton.

The low-lying parts of this stretch of coast-line are less spectacular but of great interest, both as physical features and for historical reasons. Around the Isle of Thanet is a continuous marshy area drained to the north by the North Stream and the Wantsum, and eastwards by the Stour. The towers of the old church at Reculver mark the western limit of our region on the Thames estuary. This was a former Roman station—Regulbium—guarding the northern entrance of the former strait. Near the eastern end there was formerly an island on which Portus Rutupis (Richborough) was founded. In those times there was a continuous and open strait around Thanet.

The northern entrance silted up a little, but was finally closed by walls at the end of the eighteenth century. The southern channel, although silting started earlier there, was still used by small ships up to the end of the fifteenth century, and by boats carrying coal to Canterbury until 1699. The Stour reaches the sea by a devious course. It is first diverted south by a long bank of shingle (now much destroyed and obliterated) on which Stonar stands. Then it is turned northwards by the modern beach and

enters the sea in Pegwell Bay. The Stonar shingle ridge needs explanation. Various views are held about its origin, but none is satisfactory. It is almost certain that the beach had formed in Roman times. Sandwich is a later foundation. It is also uncertain when the outer shingle began to obstruct the river. There is scope for some interesting research work on these parts of the Kent coast.

Dungeness is the greatest shingle foreland in Britain. About 98 per cent consists of flint. The photograph shows that it is formed of a great number of individual shingle ridges, each one of which at the time of its formation represented the seaward edge of the foreland. The map, fig. 1, also shows that some ridges, near the point, run parallel to the outline of the foreland whereas those farther west run at right angles to the present coast. This implies that these ridges have been largely destroyed by the sea, and also that when they were built the outline of the foreland was unlike that at present. It is, of course, impossible to trace exactly the former course of these ridges, but Fig. 1 indicates the view held by W. V. Lewis, who has studied them. It will be apparent from that figure that the point has always tended to become more and more pronounced. It seems probable that at one time there was a continuous ridge running north-east and south-west through Lydd, thus forming a great bar across the inlet which later silted up to become Romney Marsh. Fig. 1 also shows that the ridges fall into fairly well-marked groups and it is noteworthy that the average height of the groups varies considerably—differences up to eight feet are noted. This may mean that when the highest ones were formed, the level of the sea relative to the land was a little above its present level.

If it is remembered (1) that each ridge was originally the sea-line, (2) that if one ridge cuts across another, the one truncated is the older, and (3) that the point of Dungeness seems to have become more and more angular as it developed, it is possible to work out the order of formation of the ridges and to a large extent the evolution of the foreland. But many problems remain. There are a number of old shingle ridges near

Hythe. It is not at all clear just how they relate to the main mass. To the west of the foreland, the systems of ridges near Winchelsea and Camber Castle are largely independent and of later origin.

The Rother now mouths at Rye. At one time it entered the sea at New Romney. It has been argued that a former mouth was at Hythe, but this may only have been the mouth of a salt-water creek.

the Point). Nearly all the land south of the Rhee wall was reclaimed by 1500, much of it considerably earlier.

The third of the marsh and shingle spreads is that known as Langney Point or the Crumbles. It consists of a series of ridges, some of them about one mile long, running south-west and north-east. Recent changes can be traced from maps, and also by measuring the distance of the Martello towers from the sea-line.[1] It seems

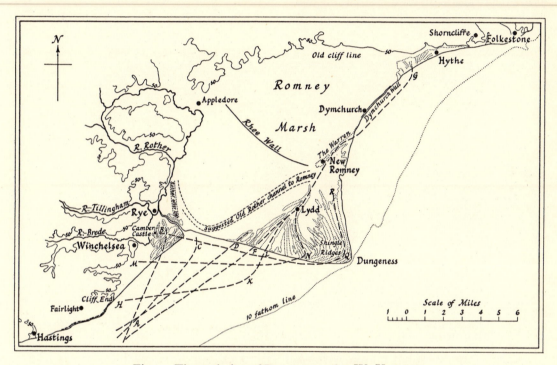

Fig. 1. The evolution of Dungeness, after W. V. Lewis.

The cliffs at Hastings extend eastwards to Pett Level and then can be traced inside the shingle and marsh. At Winchelsea and Rye they are conspicuous and it is easy to follow them past Appledore to Hythe and Folkestone, as well as round the former island of Oxney.

The history of Romney Marsh can be traced by the date of known reclamations of land. But there is uncertainty about the date at which the Rhee wall was built. It is usually assumed to be Roman; some authorities associate it with the Belgae. The earliest innings were made in the eighth century; these are in Denge Marsh (near

likely that the whole formation is fairly recent in origin—it may well be entirely post-1500. It has grown to the north-east and in so doing has blocked back the former drainage of Pevensey Levels. These correspond to Romney Marsh, and fill up an old inlet of the sea. There is no evidence of any reclamation in the Roman period. Probably the mouth was flooded at the time of the Domesday survey and a good deal of inning had been done by the end of the eleventh century. It is difficult to appreciate the changes that have taken place in the past when the ill-conceived bungalow development of the present day is contemplated!

[1] These towers were built as defences against the threatened Napoleonic invasion.

## BRIGHTON TO TORQUAY AND START POINT

### (PLATES 12–41)

This long stretch of coast cannot be regarded as a unit, but is taken for convenience' sake and divided into five sections.

(1) The first and simplest section is the Sussex coastal plain. For this Brighton forms a good eastern limit; westwards there is no clear boundary, but a rough one may be taken at approximately Chichester. The inner edge of the plain

time (1870) the opening was opposite Portslade. When the Sussex valleys were cut the land stood rather higher than now, and a later submergence has produced the present valley forms; at an earlier stage, before the accumulation of alluvium in them, there were long narrow inlets of the sea.

There have been many changes at Selsey.

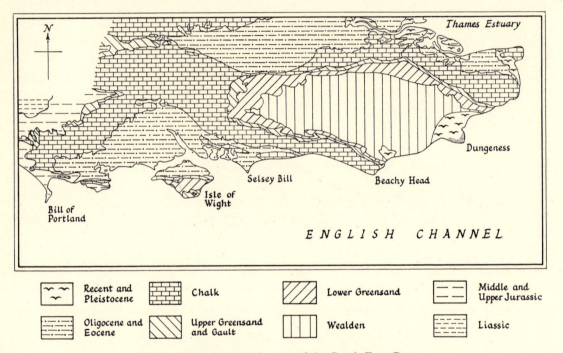

Fig. 2. Geological sketch-map of the South-East Coast.

coincides generally with the foot of the South Downs, but it is not a sharp one since it is usually hidden beneath superficial deposits. The plain widens westwards and is cut in Chalk and Eocene strata planed down to a few feet above present sea-level. It is covered by rubble and raised beach deposits.

Perhaps the most striking coastal features are the raised beaches, and the spits of shingle which deflect the rivers to the east. The changes at Shoreham harbour, now held in position by harbour works, have been considerable. At one

Bede described Selsey as an island except for a connecting shingle causeway. There is probably no truth in the view that Pagham Harbour was formed by an irruption of the sea in the early fourteenth century. Erosion is serious at Bracklesham Bay. There are very few outcrops of hard rock anywhere on this piece of coast apart from some small ones at Bognor and Barn.

(2) Chichester to White Nothe, a few miles east of Weymouth, may be regarded as one large though complicated unit. At one time the Isle of Purbeck and the Isle of Wight were connected

by a Chalk ridge, and the Frome and Stour flowed through Poole Harbour, along the north side of this former ridge, through the Solent and Spithead, and a little farther east probably turned southwards to meet a main west-flowing river along the Channel. All the rivers now running southwards, for example, Avon, Itchen, Test, and many smaller ones, and the north-flowing streams of the Isle of Wight were then tributaries to the Frome. At that time, too, the ridge running westwards from Culver Cliff to the Needles finds its direct continuation in that beginning at Old Harry Point and continuing inland behind Lulworth Cove to White Nothe.

From Chichester to Swanage the mainland coastal rocks are all soft and yield readily to erosion which locally is serious, especially between Christchurch and Milford on Sea. West of Cowes in the Isle of Wight the soft beds are

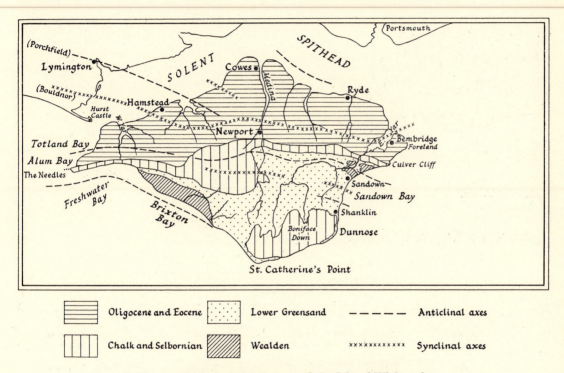

Fig. 3. Geological sketch-map of the Isle of Wight, after
H. J. O. White, *Memoirs of the Geological Survey*.

areas now occupied by Portsmouth Harbour, Langstone Harbour, and Chichester Harbour were low-lying ground drained by south-flowing streams. The Purbeck-Wight ridge was eroded partly by ordinary river action, and, on its south side, by the sea. When submergence occurred it was soon worn away, and the Isle of Wight became separated from the mainland. Southampton Water, Poole Harbour, and other inlets were formed at the same time. The old Frome made its course along soft Tertiary rocks folded in a syncline; the Chalk is deep under the Solent and reappears both to north and south. Today

subject to a good deal of slipping, and the Medina and Newton rivers, like the Beaulieu and Lymington rivers, are typically drowned forms.

The finest scenery along the coast is in Purbeck and the Isle of Wight. A brief glance at Figs. 3 and 4 shows the general arrangement of the rocks in plan and section. The rock type changes rapidly in both the isles, so that there is a great contrast between, for example, the Chalk, Greensand, and Wealden clays in the Isle of Wight. Visitors will also be familiar with the undercliff, a land-slipped area, between Niton and Ventnor, and also with the brightly coloured Tertiary sands standing in vertical beds at Alum Bay. The Isle of Purbeck is in many ways similar, but the structure has preserved the Jurassic rocks (mainly limestones and some clays and sands) along the shoreline. Hence, after leaving the Chalk at Swanage, a walk westwards along the coast shows fine cliff scenery in the Purbeck and Portland beds. Beyond St Alban's Head the Kimmeridge Clay is the main rock, and in Kimmeridge Bay the stratification is beautifully

The Bournemouth and Christchurch districts are very different. The cliffs are cut in soft Tertiary rocks, and in many places are either heavily protected or subject to serious erosion. The chines (small valleys) near Bournemouth are interesting. They show a deep inner valley cut in a broader upper one: Branksome Chine is an example. They are, in some respects, like Shanklin Chine and others in the Isle of Wight. The Bournemouth chines probably owe their

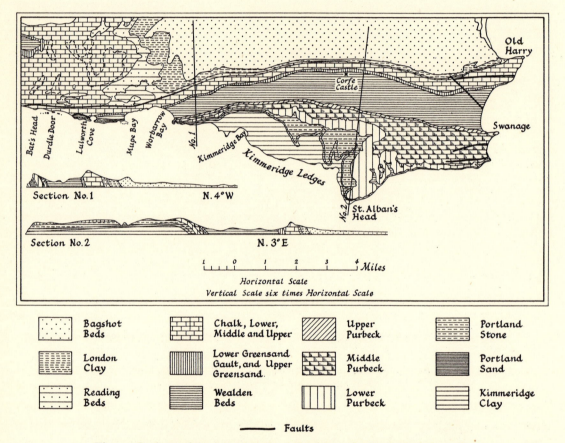

Fig. 4. Geology of the Isle of Purbeck (based on the Geological Survey).

shown. Still farther west the Jurassic rocks remain as offshore rocks or as a narrow fringe between Mupe Bay, Lulworth Cove, and Durdle Door. The sea, aided by the submergence already mentioned, has cut back to the Chalk, and Stair Hole, Lulworth Cove, and Mupe-Worbarrow Bay may be regarded as three stages. Beyond Durdle Door, a natural arch, the Chalk is the main cliff-former.

origin in the first place to meltwater from snow which covered this part in the winters of the Ice Age; the deep inner valleys were cut by small tributaries of the Solent River.

Other noteworthy features of this part of the coast include the spits of sand and shingle. Two of them enclose Poole Harbour and there has been no convincing reason put forward why one should grow from the south and the other from the north, and why the southern one should stand farther seaward. Another spit runs from Hengistbury Head, and deflects the Avon. Sometimes the spit extends as far as Highcliffe Castle. Near Milford on Sea is Hurst Castle Spit, a ridge of shingle, with several recurved ends, extending almost half-way across the Solent. Calshot Spit is smaller and situated at the mouth of Southampton Water. Behind Hurst Castle Spit, along the Solent and Southampton Water, and particularly in Poole Harbour there is an extensive development of salt marsh. The south coast marshes differ from those on the west and east coasts in two main ways. They are formed largely of soft mud, and rice grass (*Spartina*) covers them, often to the exclusion of all other plants. This particular grass is a hybrid and made its first appearance in 1870. The amount of ground covered by it since then is the clearest evidence of its power to spread and to accrete mud.

The dunes and heaths in Studland Bay (under natural conditions heathland extends all round Poole Harbour) are of great botanical interest and the seaward growth of the Studland peninsula can be traced in the main lines of dunes and the remains of former inlets, such as Little Sea.

(3) Between White Nothe and Bridport (West Bay) there are three particularly interesting shingle formations.

Before discussing these, however, a word is necessary about the cliffs between White Nothe and Lodmoor. They are formed of Jurassic rocks, mainly clays and sands with some thin limestone. They are locally subject to slipping, and near Holesworth House the rapid oxidation of iron pyrites has on occasions given rise to smouldering fires by igniting the oil shales. From 1826 smouldering continued for about four years.

The Isle of Portland is formed of Jurassic rocks, largely of limestones. There are cliffs all round it, and near the Bill there is an excellent example of raised beach. Locally the cliffs are liable to slip, especially on the coast, owing to the falling forwards of great blocks of limestone which break away along master joints.

The Jurassic rocks also form the mainland

behind Weymouth, and can be seen in section along the inner side of the Fleet. Farther north they reach the beach, and near Bridport form conspicuous vertical cliffs of Bridport Sands, capped by oolite limestone.

The first of the three beaches is at Weymouth, and is about two miles long. It is a true storm beach of large pebbles, and shuts off the low ground of Lodmoor. Radipole Lake is enclosed in part artificially. Weymouth Beach has suffered depletion since the building of Portland breakwater.

The second beach runs from Portland Castle to Small Mouth. In a part of its course it touches the Chesil Beach but is quite distinct from it. It is about one and a half miles long, encloses the Mere, and is built mainly of Portland and Purbeck rocks.

The Chesil Beach is unique in these islands, if not in the world. It runs from Bridport to Portland island, and south-east of Abbotsbury it is separated from the land by the Fleet. It is about 170 yards wide at Abbotsbury, and 200 yards at Portland. Its height at the same two places, measured from high-water mark, is nearly twenty-three feet and nearly forty-three feet. The gradient of its crest-line averages about 1 in 8500 from Abbotsbury to Wyke, and 1 in 880 from Wyke to Chesilton. At Abbotsbury the shingle extends some six fathoms downwards below low-water springs and at Portland eight fathoms. The stones above sea-level increase in size with marked regularity from north-west to south-east; at Bridport they are about the size of a hazel-nut; at Portland perhaps as large as a goose's egg. Below low-water the size is said to decrease in the same direction.

The beach is made up of 90 per cent flint pebbles, and others from the Portland and Purbeck rocks, as well as from more distant sources, including the rocks found in Devon and Cornwall. Many may be redistributed from raised beaches which at one time certainly existed hereabouts: that at Portland Bill alone remains. It has been, and still is, a puzzle to explain the formation of the beach. It seems to be fed from both ends; the rocks derived from Portland Island can be traced some way along the beach, and there is good reason to think that there is a definite movement eastwards from Bridport. Granting this, it is still more difficult to explain the very even grading of the pebbles, and it is in this respect that the Chesil is unique. Experiments have shown that if, for example, brick fragments are thrown down on the beach, they will move to that part of it where there are pebbles of similar size.

There is no doubt that waves are the main agent in the formation of this beach. In great storms it is still occasionally overtopped. It has joined Portland to the mainland and the Fleet is a drowned feature, the former right bank of which has disappeared and been replaced by the Chesil. Many distinguished people have written about the origin of the beach, but the complete explanation is still awaited.

(4) Westwards from Bridport to Babbacombe (Torquay) the coast is fronted by cliffs of Jurassic, Triassic, and Permian rocks. Many of the rocks are highly coloured—for example the dark grey and black limestones and clays, sometimes capped by yellow sands, near Lyme Regis, and the bright and dull browns and reds at Sidmouth and Torquay.

Black Ven, Stonebarrow, Golden Cap, and Thorncombe Beacon form conspicuous hills and high cliffs between Lyme Regis and Bridport. The beds in them, mostly horizontal or but very gently inclined as seen from the sea, vary greatly. The clays often lead to slips, and erosion is locally severe. The Upper Greensand gives a bright coloured top to Golden Cap (hence the name), and the four main hills are separated by small valleys: the Char, St Gabriel's Mouth, Seatown, and Eype Mouth. The view from the Cobb at Lyme Regis is magnificent. The Cobb itself stands on limestone, and the synclinal arrangement of the beds can be clearly seen.

Westwards from Lyme Regis the dark cliffs of alternating thin beds of clay and muddy limestone soon give place to one of the most interesting features on the British coasts, the great landslip near Rousden. The Cretaceous rocks (Chalk, Greensand, Gault Clay) rest on mainly clay rocks of the Jurassic series with a fairly

marked break. Between Culverhole Point and Humble Point the inland cliffs are of Chalk and Upper Greensand; the undercliffs of fallen Cretaceous material much disturbed. They slope down eventually to rest on the Lias shore reefs. Dowlands' Chasm, formed in 1839, is the most important feature. From June 1839 there had been a period of much rain and strong gales. Fissures and cracks began to appear on the cliff-top shortly before Christmas 1839 and on 23 December one of the cottages began to subside, at first in no very alarming manner. But by 5 p.m. the cottage was settling rapidly, and later other cottages were 'upheaved and twisted'. The great landslip occurred on Christmas night. 'During December 26 the land that had been cut off by the fissures in the cliff top gradually subsided seawards. A new inland cliff, 210 feet high in its central position and sinking to east and west, had thus been exposed, backing a chasm into which some twenty acres of land had subsided. The length of the chasm was about half a mile, while its breadth increased from 200 feet in the west to 400 feet in the east.' The ash forest has grown naturally since the slip, and the bare scars have, of course, all disappeared. A walk through the slip today is most revealing, and but little imagination is required to picture the landscape as it was at Christmas 1839.

There are other landslips in this neighbourhood, but on a smaller scale. That at Hooken Cliff, between Branscombe and Beer Head, is perhaps the best known.

Beyond Branscombe the coast is very beautiful. The cliffs are high and deep red in colour. There are several deeply cut combes which diversify the cliff line, and Sidmouth stands at the mouth of the biggest one. Two miles farther west, in Ladram Bay, the sea has fretted out an interesting series of stacks. The site of Budleigh Salterton is not unlike that of Sidmouth, and the natural beach, unspoilt by a promenade, is attractive. All the way to Babbacombe the Permian rocks form the cliffs and red and red-brown colours prevail. It is a well-known coast that includes Exmouth, Dawlish, and Teignmouth. Moreover, since the main line to Plymouth runs close to the sea for some miles it is

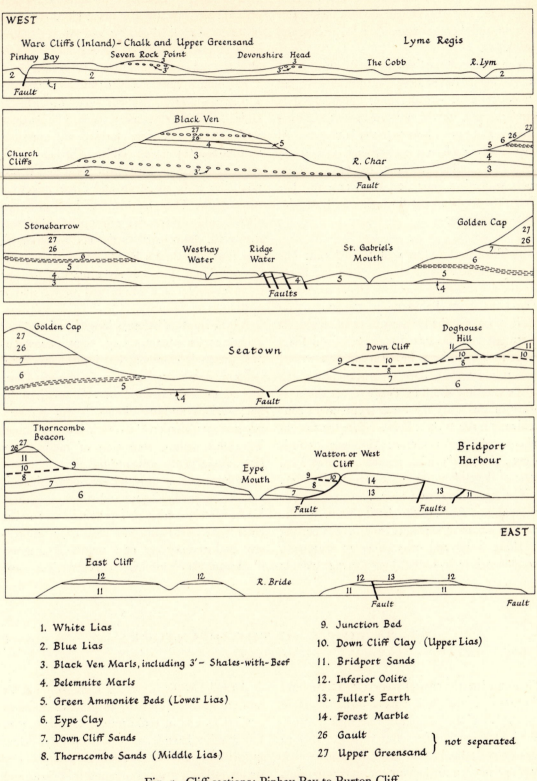

1. White Lias
2. Blue Lias
3. Black Ven Marls, including 3' - Shales-with-Beef
4. Belemnite Marls
5. Green Ammonite Beds (Lower Lias)
6. Eype Clay
7. Down Cliff Sands
8. Thorncombe Sands (Middle Lias)

9. Junction Bed
10. Down Cliff Clay (Upper Lias)
11. Bridport Sands
12. Inferior Oolite
13. Fuller's Earth
14. Forest Marble
26. Gault } not separated
27. Upper Greensand }

Fig. 5. Cliff sections: Pinhay Bay to Burton Cliff,
after G. M. Davies, *The Dorset Coast*.

also known in general terms to many who do not stop there. The cliffs are subject to considerable erosion, and there are several stacks, arches, and headlands.

The mouths of the Exe and Teign are drowned valleys like those we have met farther east, but are not yet filled with alluvium. The tidal waters run far inland, and the alternate exposure of sand and mud add great beauty to the landscape. At the mouth of each river is a sand bar. That across the Exe is double; the outer bar has of recent years suffered a good deal from erosion. The Teign bar seems to run through a fairly regular cycle of changes.

(5) Tor Bay is a fairly well-defined unit. The enclosing headlands of Hope's Nose and Berry Head have resistant cores of Devonian rocks. The coast from Babbacombe to Start Point really belongs to the Devon-Cornwall region. It is treated here because Lyme Bay from Start Point to Portland can be regarded as a whole, and, although old (Palaeozoic) rocks form much of the coastline south of Torquay, for the purposes of this book the break seems more conveniently made at Start Point, where the ancient metamorphic rocks outcrop. Moreover, there is a good deal of Permian faulted down to form most of the cliffs in the inner parts of Tor Bay.

The structure at and near Torquay is complicated, and much of the detail that makes the coast so picturesque results from the sea cutting in along faults and other lines of weakness. London Bridge is eroded along cleavage planes;

Hope's Nose is a remnant of limestone resting on a thrust-plane (fault); Black Head is a mass of volcanic rock (dolerite); and Anstey's Cove owes its origin to the sea having cut into slates after the removal of the limestone cover.

Paignton is on the Permian; and between there and Brixham sand, grits and shales occur in the cliffs. Broad Sands rests on slates. The Brixham promontory is mainly formed of limestone surrounded by steep cliffs. Farther south, especially between Sharkham Point and Blackpool, the cliffs are steep and often somewhat inaccessible. They are made of hard grits and shales, and differential erosion is admirably displayed. Igneous rocks also add much detail, and form Matthew's Point, Redlap Cave, and Coombe Point. The drowned estuary of the Dart is an example of a submerged valley, and is outstandingly beautiful.

A little south of Strete, a long bar of sand and shingle encloses Slapton Ley. Flint is conspicuous, quartz is common, but immediately local rocks are poorly represented in the shingle. There is room for some research into the origin of the bar; various authors discuss it but no very adequate explanation of its origin is forthcoming. The land within, consisting of reddish-grey slate, slopes gently down to the ley.

At Beesands, a little farther south, there is another but smaller lagoon and at Hallsands there has been serious erosion. The general view of the coast from Start Point on a clear day is impressive and reminiscent of a text-book diagram illustrating the simplification of a stretch of coast.

## START POINT TO THE QUANTOCKS
### (PLATES 42-69)

To appreciate the coastal scenery of Devon and Cornwall to the full entails a considerable knowledge of the complicated geological structure of those counties. But it is possible to make some rather broad generalizations which may be sufficient to illustrate the more important features. In this peninsula there are several high moorlands. Those of Dartmoor, Bodmin Moor,

St Austell Downs, Carn Menellis, and Land's End are formed primarily of granite. Exmoor consists mainly of sediments of Devonian age. The granites have been pushed up from below and are now exposed as a result of erosion. They are all within a great compound fold (syncline) the axis of which runs roughly east and west. The Culm Measures form the heart of this fold. In

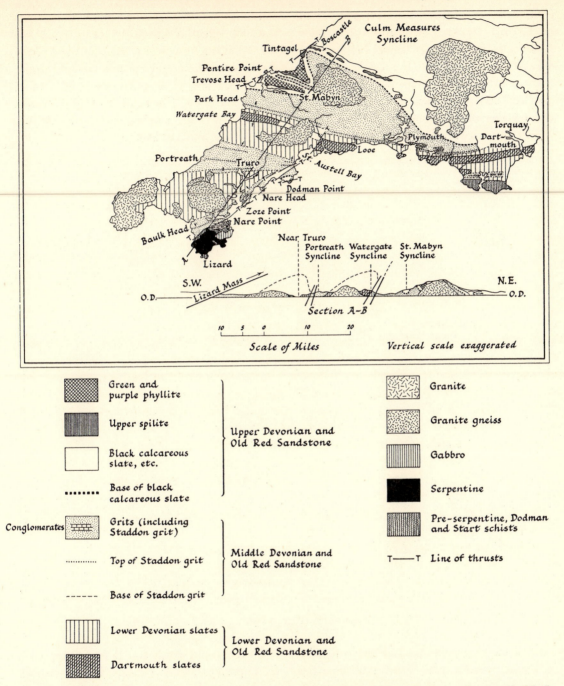

Fig. 6. Geological sketch-map of Devon and Cornwall and section through Cornwall, after E. M. L. Hendriks, 'Rock Succession and Structure in South Cornwall', *Quart. J. Geol. Soc. Lond.* **93** (1937), 322.

the south there is a complicated upfold, the trend of which again runs east and west and meets the Atlantic in Watergate Bay. It includes many different rocks, grits, slates, and sandstones. In the far south near the Lizard and also in the Prawle Point–Start Point districts of Devon are the oldest rocks of all, the serpentines and the schists. The rocks between them and the Culm Measures consist of a variable series of sandstones and slates striking mainly east and west.

It follows that despite many local folds and

faults, which often give rise to interesting details on the coast, there is a rough correspondence of rocks on the Atlantic and Channel coasts. Reference to Fig. 6 shows this. But it does not follow that the two coasts resemble one another very closely. The Atlantic coast is far more exposed and for this reason alone the cliffs are often steeper and more imposing. Along the Channel

The varying hardness of the many different beds, the effect of igneous dykes intruded into the sediments, the great number of minor folds in the rocks, the deeply cut valleys, and the fact that the sea now penetrates in many inlets of all shapes and sizes, give to the Cornubian coasts a variety of detail matched only by Pembrokeshire in England and Wales.

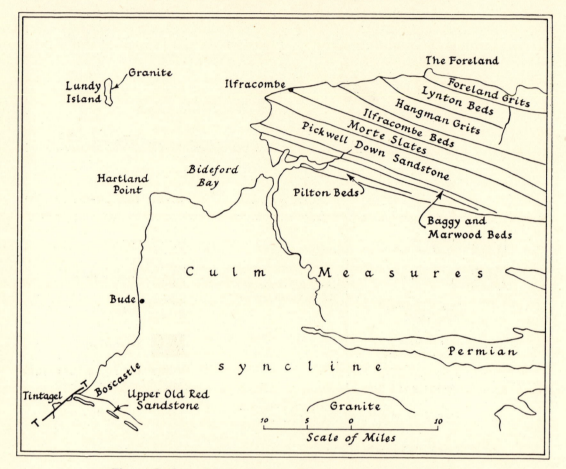

Fig. 7. Geology of the north coast of Devon, after E. A. N. Arber.

coast of the peninsula there are many remains of a raised beach which shows itself as a low platform about ten feet above high-water mark. On the north coast, there are traces of high-level platforms which give a characteristic appearance to the cliffs—Reskajeage Downs illustrate this remarkably well. In other parts, especially in Devon, many of the cliffs owe but little of their form to marine action; they are really steep slopes produced by sub-aerial erosion, and only their 'toes' are modified by the waves.

Between Start Point and Bolt Tail the cliffs are imposing; locally the raised beach is well developed, and its cover of 'Head' is conspicuous. Salcombe Harbour and Plymouth Sound are two good examples of the type of drowned inlets called rias. There are many other beautiful inlets—Erme Mouth, Yealm Mouth, Looe River, Fowey River, and Carrick Roads. Between Plymouth Sound and Mevagissey, slates, often of bright colours, are the main cliff-formers. Farther west there are many fine headlands, nearly all of which are made of harder rocks. The numerous picturesque coves are cut in the softer rocks. Many examples could be given, and the following may be regarded as typical. Resistant conglomerates enclose Portloe Cove, Kiberick Cove lies between the igneous masses of Nare Head and the Blouth, the Straythe between the Blouth and Manare Point. Jacka Point, Hartritza Point, and Caragloose Point are all formed of igneous rocks. It is not, however, always easy to say why a headland is so prominent. The Dodman, for example, is a uniform mass of phyllite. The cleavage of the rock is steeply inclined to the south-east, and a good deal of slipping occurs. On the western side the cliffs are steep. Nevertheless, the rocks of which it is composed are not particularly resistant.

Similar features are found farther west. Near Pendennis and between Pennance Head and Maen Porth there are several narrow gullies, called drangs, which have been eroded along small faults. The raised beach is conspicuous and sometimes there are old caves at the back of it. The many beautiful sand beaches at the head of the bays and inlets are a noteworthy feature of all the south coast of Devon and Cornwall.

The Lizard district has some magnificent cliffs cut in schists (for instance, Old Lizard Head). The most characteristic rock is the serpentine which reaches the sea at Kennack and Coverack, between Mullion and Kynance, and from the Lizard to Cadgwith. It is cut by many dykes of igneous rock, which often give rise to interesting features such as Carrick Luz. The Lizard rocks are hard and stand out, as can be seen at Porthhallow and Polurrian. The interior of this part of Cornwall is smooth and bare, so that flat-topped cliffs are characteristic.

In Mounts Bay, Loe Bar is a shingle ridge enclosing Loe Pool. Like Prah Sands it is mainly formed of flint shingle. Because flint is derived primarily from the Chalk, and since there is no Chalk outcrop nearer than eastern Devon, the occurrence of flint in these beaches has given rise to various hypotheses. St Michael's Mount is a granite hill and was once part of the main-land; there is no known record of the time when it became an island.

The Isles of Scilly, about 30 miles south-west of Land's End, are all granite, and represent the remains, cut up both by sub-aerial and by marine erosion, of a former granite mass like that of the Land's End peninsula. A noticeable characteristic of the Scillies is the number of sand and shingle bars and impounded ponds of salt water. St Mary's Island, for example, is made up of three distinct islands joined by bars. There is evidence of subsidence in these islands, which together with the rocks called the Seven Sisters at one time stood higher and probably formed an area of dry land—Lyonesse.

Land's End, a granite cape, is the most westerly point of Britain. Although the cliffs are fine, they are not equal to those near Penberth which show the weathering of granite into great castellated masses extremely well. Moreover, the cliff-top of open moorland at Penberth is superb. North and east from Land's End the granite forms many cliffs, which are often interrupted by coves. The cliff slope is locally rather gentle. Many narrow inlets—zawns—are the result of the removal of vein material of tin or copper lodes which was softer than the enclosing rocks. There are numerous remains of former mining along this part of the coast.

In St Ives Bay there are some high dunes. Dunes also occur in other parts of Cornwall, and in some places they have advanced inland and partially buried churches and other buildings. At Perranzabuloe (=Sanctus Piranus in Sabulo) the oratory of St Piran was overwhelmed in this way. The Gwithian Sands invaded Lelant Church, and St Constantine's Chapel was also destroyed. It is relevant to add that a great deal of legend attaches to some of these places and obscures facts which are usually more sober!

Between St Ives and Newquay the most prominent features are St Agnes's Beacon and St Agnes's Head, formed of much-altered slates. South of the head differential erosion is displayed in many coastal features, stacks, and islands—Gull Rock, Samphire Island, and Crane Island, for example. There is also a remarkably fine line of cliffs at Reskajeage and Hudder Downs. North of St Agnes is Perran Bay with its fine beach. Here, as well as in many smaller bays and inlets, are dunes; sometimes there may be a small impounded lake. At Perranporth and Padstow the sand is mainly shell sand and very light, and is easily moved by the wind. At and near Newquay the coastal rocks are much broken by faults, and many caves are eroded along them. Differential erosion is evident: the Criggars and Towan Head are rocky ledges of silicified shales and limestones; East and West Pentire Heads are made of hard rock, whereas the shale cliffs in Fistral Bay weather rapidly. The Gannel is a drowned valley, and may be compared with Watermouth near Ilfracombe. The Camel estuary is a drowned mouth partly obstructed by the Doom Bar. In Watergate Bay, black shales form cliffs about 200 feet high, and the crest of the anticline (page 119) is near Horse Rock.

North of the Camel the prominent features are nearly all made of hard igneous rocks—Stepper Point, Gulland Rock, the Mouls, Rumps Point, Cliff Castle. The bays are generally cut in softer shales. At the north end of Port Isaac Bay there begins perhaps the finest stretch of cliffed coast in England and Wales. Tintagel is complex in structure; the island is a flat-topped piece of the interior plateau separated by erosion which has, in fact, followed lines of faulting. The cliffs, islands, and stacks to the north of Tintagel are bold, and it is a great pity that some cliff-top buildings in this neighbourhood mar the natural beauty of the sky-line.

A short and adequate description of the scenery from Tintagel to Hartland Point is impossible. The coast must be seen to be appreciated. South of Bude the folding in the cliffs produces innumerable ledges and small re-entrants, and the coastal platform shows similar features. The harbour at Boscastle is a good example of a submerged river mouth. North of Bude the cliffs are magnificent and the numerous coastal waterfalls are of great beauty and interest. Sometimes the fall is vertical (Litter Water), others break up into distinct parts, and their courses depend upon the dip and strike of the rocks. Milford Water can be divided into six sections. Sometimes the former streams ran for a short distance roughly parallel to the coast, and then spilled over a cliff. The old course of Wargery Water behind both Hartland and Screda Points is easily traced on the ground. Hartland Point is a right-angled turn in the coast. Eastwards from it the cliffs, like those to the south mainly of sandstones and shales, are steep and flat-topped. Clovelly street was once a natural watercourse. Farther east the cliff profile changes, and is of the hog's-back type. Nearer Appledore the most conspicuous feature is the coastal platform cut in much-folded rocks. Seen from the cliff-top it resembles an exercise in geological mapping!

Barnstaple or Bideford Bay, including the combined mouths of the Taw and Torridge, marks a break in the coast. On its south side is the pebble ridge, or Popple, of Westward Ho!, and on the north the fine beach and dunes of Braunton. These give place to an idented coast with fine bays and beaches, Croyde Bay and Woolacombe Sands, separated by a prominent headland. Between Combe Martin and Lee Bay (near Lynton) there are some typical hog's-back cliffs. Much of the coast is heavily forested and its beauty depends greatly on the fine country inland, such as that near Trentishoe and Heddon's Mouth. The Two Hangmans, Holdstone Down, and Trentishoe Burrows result from the differential weathering of a table-land. Near Lynton the Valley of the Rocks is an old, and now dry, river valley, the seaward wall of which has been breached. Hog's-back cliffs continue as far as Porlock, where the eastward travel of shingle has given rise to a small foreland around the weir and harbour. It is, however, the combination of coast and interior—Exmoor—that gives such beauty to this part of the country.

Beyond Porlock is the high ground between Hurlstone Point and Minehead. Hereabouts the harbours are often in soft rocks between fault blocks. The coast east of Minehead is lower, but the picturesque harbour at Lilstock, the waterfall in St Audrie's Bay, and the fine setting of the coast and the Quantocks are noteworthy.

## THE SEVERN ESTUARY
(PLATES 70–72, AND SEE FIG. 8)

Upstream from the Parret on the south and the Taf on the north the Bristol Channel narrows, and becomes a true estuary. There are, nevertheless, several coastal features of interest from the point of view of the physiographer.

The great tidal range, about forty feet in places, and the large amount of mud carried by the tidal waters are noteworthy. The shores of the estuary are usually low and flat, and wide mud-flats are locally exposed at low water. On the south side there are several areas of marsh and fenland, and the Parret marshes make a close parallel with the East Anglian fens. The Somerset levels are, however, much cut up by the high ground of the southern Cotswolds, the Mendips, the Poldens, and the Quantocks. Away from the river mouths the coast is lined with dunes, especially between Burnham and Brean, where the belt may be three-quarters of a mile wide. Salt marshes are mainly restricted to the river mouths.

Some interesting changes have taken place at the mouth of the Parret. This is deflected northwards by a shingle spit, which has already incorporated the former island of Fenning and is growing towards Steart Island. Farther north the headlands on the east, with the exception of the Cleveland–Portishead ridge, were formerly islands. Brean Down, Worle Hill, and Middle Hope are tied to the mainland by sand-spits. Brent Knoll is now a long way within the marshes. Steep Holme and Flat Holme are limestone islands.

Sedbury and Aust cliffs were once continuous. Aust cliff is a ridge of Triassic and Lower Jurassic rocks which appear to have been hardened either by folding or faulting or both so that they have resisted erosion. The colours as well as the steepness of the cliffs are remarkable.

Nearby is the drowned mouth of the Wye, and a little farther south and west are the Caldicot and Wentloog Levels, both of which are formed of bluish clay of tidal origin. Today they are defended by a sea-wall outside which are the bare mud-flats. The contrast between Cardiff and Newport and the rural, even remote, nature of parts of the Levels is striking.

## SOUTH WALES, CARDIFF TO ST BRIDE'S BAY
(PLATES 73–88)

In this interesting length of coastline, a primary distinction must be made between south Glamorganshire and the remainder. The Glamorgan coastal rocks are mainly the Trias, Rhaetic, and Lias. On the whole they are less resistant than the older rocks farther west. Nevertheless locally they give a remarkably beautiful coast, especially at Southerndown, where the foreshore rocks are Carboniferous overlain by the Sutton Stone of the Lias, and at Nash Point. Farther east many faults cut the coast, and are partly responsible for small bays such as St Donat's and Stradling Well. Around the mouth of the Thaw (Ddaw) there is much shingle, blown sand, and alluvium. Lavernock Point, where the coast turns north, may be regarded as the eastern limit of the Bristol Channel.

Much of the coast west of Southerndown is bordered by extensive dunes, but as it will be more convenient to discuss them in a later paragraph, we may first turn to the Gower peninsula. Fig. 9 shows the arrangement of the rocks. It will be apparent that folding, trending nearly east and west, has caused much of the coast to be fronted by rocks of Carboniferous Limestone, but in Oxwich and Port Eynon Bays there is Millstone Grit, and in Rhossili Bay some Old Red Sandstone, which inland forms the highest ground of the peninsula. Fig. 9 also shows that there are many lines of faulting, and

123

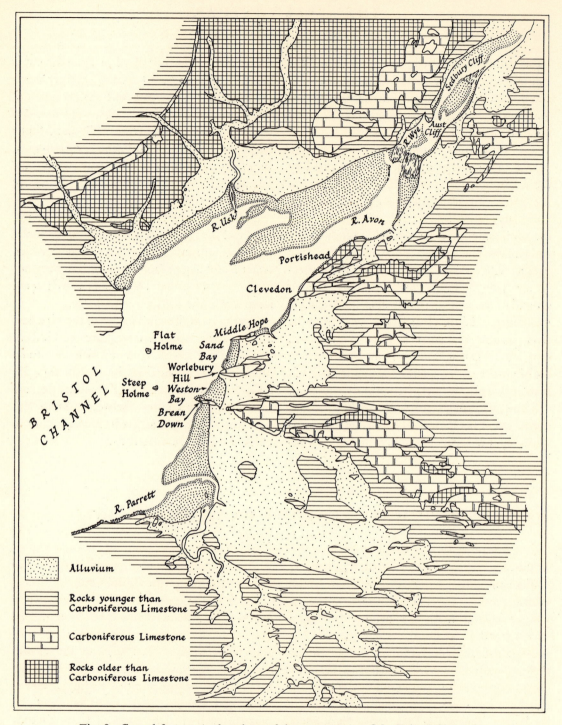

Fig. 8. Coastal features and geology of the upper parts of the Bristol Channel, based on the Geological Survey.

that in several cases the faults cut the coast at a high angle. Faults, and erosion working along them are largely responsible for Caswell Bay, Brandy Cove, Pwll-du Bay, and Three Cliffs Bay. On the other hand, Swansea Bay and the big inlet of the Burry River are in the softer rocks of Upper Carboniferous age.

The Gower cliffs are beautiful, the finest range being between Worm's Head and Port Eynon. They are high and cut into coves and headlands. A low platform along the foot of these cliffs represents the remains of a raised beach, called the *Patella* beach after the limpet which is commonly found in it. Another platform, in Rhossili Bay, at a level of about 100 feet, that at first sight looks like a raised beach, is of uncertain origin. Port Eynon and Oxwich Bays are backed by dunes and woods. Farther east, the alternation of smaller bays and lines of good but less spectacular cliffs, and the prominence of the lower (*Patella*) platform render the coast less imposing, although attractive. The railway and other man-made features spoil the coast facing Swansea Bay.

On the north side of Gower the cliffs, still of Carboniferous Limestone, no longer face open water, but are fringed by an extensive salt marsh that occupies much of Burry inlet. The marsh plants resemble those in the Dovey (see page 129). The western end of the marsh is enclosed by a pebble ridge capped by dunes, called Whiteford Spit.

From Llanelly to Pendine there are extensive dunes, and there is only one short line of picturesque cliffs—that in the Llanstephan peninsula which is bordered by the unspoiled estuaries of the Taf and Towy. Old cliffs are conspicuous behind Laugharne Burrows.

From Amroth to Saundersfoot and Monkstone Point there are many faults and folds in the Upper Carboniferous rocks which extend from Tenby to Ragwen Point, which is formed of quartzites. Telpen Point is made of the Farewell Rock, the top of the Millstone Grit series. In this stretch shales usually coincide with bays, and sandstones with promontories, such as Monkstone Point.

The Tenby peninsula is a replica of the Gower peninsula. The Coal Measure rocks (Fig. 9) are all north of Tenby, and Silurian and Ordovician strata outcrop in but limited areas. Hence, in broad terms, the effective rocks in the production of both inland and coastal scenery are the same in the two peninsulas. The folding near Tenby is perhaps a little more complicated, but here again the trend is nearly east and west. The reader may note the similarity between these two peninsulas and that between the Isles of Purbeck and Wight on the south coast. The parallel is indeed close, and the situation of Carmarthen Bay closely resembles that of Bournemouth Bay. The difference between the rocks of the two coasts is one of age. Those on the south coast, the Chalk, Greensand, and clays, are all Mesozoic or Tertiary; those in South Wales are all Palaeozoic. The two areas typify two different periods of folding; subsequent events have produced the present outlines.

Milford Haven is the northern boundary of the Tenby peninsula, in which the disposition of the rocks has a direct effect on coastal scenery. Apart from superficial deposits, West Angle and Angle Bays correspond to Freshwater East Bay, and the Pembroke River with part of Lydstep Bay. As in Gower, the many small faults play a large part in the formation of small bays, such as Swanlake, Manorbier and Precipe. Caves, too, are usually formed along faults. There are strong contrasts between cliffs in different rocks: for example, those in the grey limestones of the Carboniferous and those in the reds and browns of the Old Red Sandstone. The contrast is well seen on the two headlands enclosing Freshwater West Bay, and even better in Freshwater East Bay and Skrinkle Haven. Moreover, because of the folding, the beds in the cliffs, no matter what the type of rock, may stand vertically as near Old Castle Head and Lydstep or horizontally as at St Gowan's Head. Sometimes the bedding is on a massive scale; in other places in relatively thin layers.

In the magnificent limestone cliffs between Linney Head and St Gowan's Head there is an infinity of detail. The Green Bridge is a natural arch, the Huntsman's Leap is a great fissure coinciding with a small fault. Much of the detail of the coast is not directly the result of marine erosion acting on the cliffs, but of marine erosion acting on a mass of limestone already riddled with caves, swallow holes, faults, and other lines of weakness. One major feature in the coastal scenery both here and at Gower, but at right angles to the trend of the coast. It owes its present appearance largely to submergence of a system of river valleys. Submergence is also partly responsible for the off-lying islands of Skokholm and Skomer. The Dale peninsula is almost an island, and separated from the main-

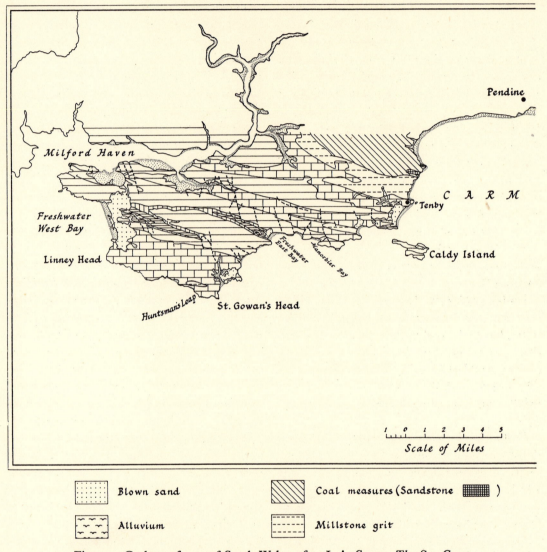

Milford Haven

Freshwater
West Bay

Linney Head

Huntsman's Leap    St. Gowan's Head

Freshwater
East Bay    Manorbier Bay

Pendine

C A R M

Tenby

Caldy Island

ı 0 ı 2 3 4 5
Scale of Miles

Blown sand      Coal measures (Sandstone     )

Alluvium        Millstone grit

Fig. 9 a. Geology of part of South Wales, after J. A. Steers, *The Sea Coast*.

unusually conspicuous between Linney and St Gowan's Head, is the flat top to the cliffs. The flat surface may be a plane cut by the sea in former times when the relative levels of land and sea were different.

Milford Haven is a ria, an inlet of the sea formed in rocks the strike of which runs roughly land by a deep valley. The coast north of St Ann's Head is fine, especially in Marloes Bay with its stacks and varied rocks. Although Skomer is partly insulated by submergence, erosion working along numerous small faults is still vigorously at work, and various stages in island formation can be seen in the Deer Park peninsula, Midland Isle, and Skomer. Gateholm is a great stack of Old Red Sandstone not quite separated from the mainland.

The Old Red Sandstone makes good and deeply coloured cliffs along part of the south side of St Bride's Bay. Often, however, cliffs are

storm-beach dam back a marsh and Bathesland Water. The stones are of local origin.

This corner of the bay is an important geological boundary. The rocks south therefrom are all more recent in age than those along the north coast of the bay. Moreover, the new rocks

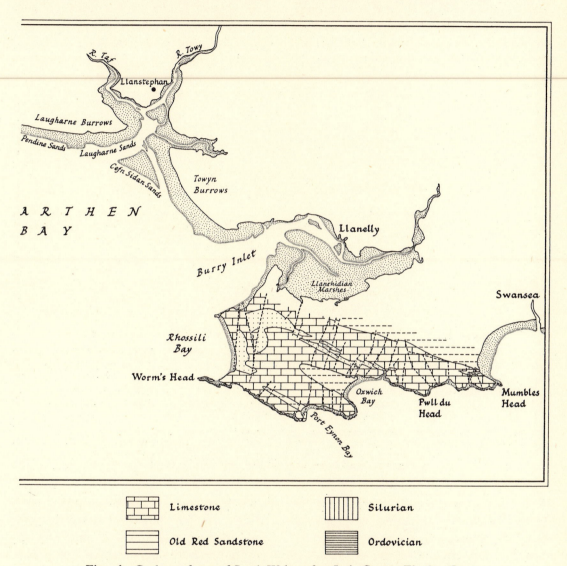

Fig. 9b. Geology of part of South Wales, after J. A. Steers, *The Sea Coast*.

cut either in igneous or in relatively soft Carboniferous rocks. The minor headlands are usually in sandstones, the small bays in shales. The detail near Broad Haven and Little Haven is interesting in relation to the folding and faulting that can be clearly seen in the cliffs. At the north-eastern corner of St Bride's Bay, Newgale Sands and

have also been folded at a later period, so that not only do two different rock groups meet at this point, but also two quite distinct periods of folding, the Hercynian (or Armorican) to the south, the Caledonian to the north. We have already stressed the importance of east–west trends in South Wales; in Cardigan Bay and

North Wales, the trend of the rocks is more nearly south-west to north-east, although in the St David's peninsula the Caledonian trend is bent round so that it is nearly east and west.

There are numerous sand-dune areas along the coast of South Wales. Nearly all occur in bays, especially in those facing a westerly direction. In general, the size of any particular mass of dunes depends upon the size of the bay and the nature of the land behind the dunes. St

[=dunes]...to the Early Iron or Late Celtic period, i.e. to a time beginning say 400 B.C. and extending up to the Roman occupation of the districts.' L. S. Higgins and others have investigated dunes in Glamorganshire, especially those between Swansea and Llanelly. Hereabouts, especially near Kenfig, the interrelation of dune movements and historical events is significant. The evidence from nearly all these dune areas is fairly consistent. In the historic period no

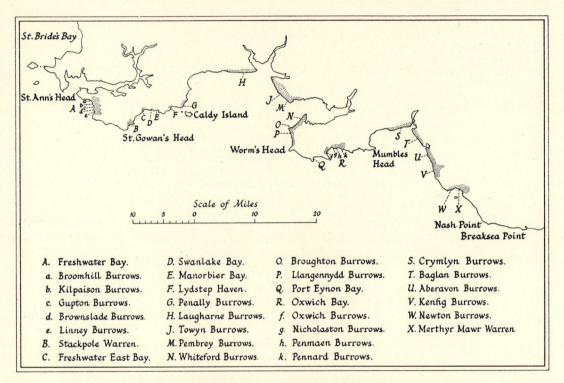

| | | | |
|---|---|---|---|
| A. Freshwater Bay. | D. Swanlake Bay. | O. Broughton Burrows. | S. Crymlyn Burrows. |
| a. Broomhill Burrows. | E. Manorbier Bay. | P. Llangennydd Burrows. | T. Baglan Burrows. |
| b. Kilpaison Burrows. | F. Lydstep Haven. | Q. Port Eynon Bay. | U. Aberavon Burrows. |
| c. Gupton Burrows. | G. Penally Burrows. | R. Oxwich Bay. | V. Kenfig Burrows. |
| d. Brownslade Burrows. | H. Laugharne Burrows. | f. Oxwich Burrows. | W. Newton Burrows. |
| e. Linney Burrows. | J. Towyn Burrows. | g. Nicholaston Burrows. | X. Merthyr Mawr Warren. |
| B. Stackpole Warren. | M. Pembrey Burrows. | h. Penmaen Burrows. | |
| C. Freshwater East Bay. | N. Whiteford Burrows. | k. Pennard Burrows. | |

Fig. 10. Sand-dune areas on the South Wales coast.

Bride's Bay is exceptional in having very few dunes. The smaller dune groups in Pembrokeshire need no particular comment. In Carmarthen Bay there are much larger expanses, and here, as well as farther east, there is often traceable a connection between the growth and inland penetration of the dunes and the destruction of property. Sometimes, too, traces of early man are found. T. C. Cantrill, writing of the part of the coast between Watchett Pill and Ginst Point, says: 'All things considered...I am disposed to refer the first definite occupation of the burrows

danger from moving sand was suspected up to the early part of the thirteenth century. In north-western Europe as a whole, the thirteenth and fourteenth centuries were times of great storms; in the fourteenth century in South Wales a comparatively sudden change occurred, the records all implying that storms blew up much sand which did great damage to property and to cultivated land. It is possible that part of the trouble may also have been caused by a slight downward movement of the land in relation to sea-level.

## WEST AND NORTH WALES

### (PLATES 89–114)

Northwards from St Bride's Bay Wales is characterized by a north-east to south-west Caledonian trend, which makes itself apparent in the Lleyn Peninsula, the coast of Cardigan Bay, and in the Teifi and Towy valleys. In Pembrokeshire, however (see p. 128), the trend has turned almost east and west.

From Newgale to Cardigan is one of the finest stretches of coast in Britain. The lichen- and plant-covered cliffs near St David's and Solva, the many picturesque inlets like the St Nons, Caerfai, and Caer Bwdy bays, and the numerous stacks and islets, nearly all closely related to the relative hardness of the different rocks of Cambrian age, are memorable. Ramsey Island is a detached part of the Pembrokeshire plateau, and a strong tide-race runs through Ramsey Sound. The outline of the island is in large part the result of wave attack on rocks of differing resistance. Nearer St David's igneous masses begin to play a more important role. The dunes in Whitesand Bay are in strong contrast to the high ground of Pen Beri and Carn Llidi. Reference to Fig. 11 shows the intricacy of the coast, and how harder rocks stand out to form headlands. All the way to Strumble Head this is wonderfully shown, and probably no part of the British coastline displays the interdependence of structure, rock hardness, exposure, and coastal features better. Even in minute detail there is a striking parallel between coastal crenulations and variations in rock type. The importance of igneous rocks is clearly brought out in Fig. 11.

The coast of Cardiganshire south of approximately New Quay really prolongs that of Pembrokeshire. The dark cliffs of shales and sandstone are interrupted by inlets such as Cwm Tydi, Llangranog, and Aber Arth (just north of New Quay). The estuaries of the Teifi and Nevern are partly blocked by beaches which anticipate the much greater spits and bars of sand and shingle farther north. It is probable that at one stage of the Ice Age, ice ponded back water in some of these inlets and produced glacial lakes, such as Lake Teifi. Perhaps the deep valley which all but makes Dinas Head an island was cut by meltwater from an ice-sheet. The detailed relations between folding, faulting caves, and coastal detail are remarkably displayed in the cliffs near Cemmaes Head.

North of New Quay the coast is simpler in outline. Boulder-clay flats and cliffs, and beaches of coarse cobbles become more conspicuous. Between Llanrhystyd and Morfa Mawr there is a well marked boulder-clay platform about half a mile wide, which slopes gently downwards from the 100-ft. contour to low cliffs faced by shingle. The boulder-clay spread is almost continuous between Aber Arth and Aberayron. At Cwm-Ceirw the cliffs, vertical and gullied, are entirely cut in boulder-clay. A little farther north Aberystwyth Grits form a steep rocky coast. The westerly dip is about 30°, and this mainly controls the slope of the cliffs. Hereabouts landslips are not uncommon, and take place on a small scale along the joints and bedding planes.

Aberystwyth might better be called Aber Rheidol, but possibly the Ystwyth once flowed nearer the town than it now does. At and near Aberystwyth erosion is often serious, and part of the northern promenade was destroyed in the great storm of 22 November 1938. North of the town, cliffs are prominent, and broken by the river Wallog and by Clarach Bay; in the former there is a good deal of boulder-clay. The cliffs continue to Borth, but north of that place the coast changes. There are very few cliffs—and not all of them face open water—between Borth and the end of the Lleyn peninsula.

North of Borth there is a great storm beach which extends almost across the estuary of the Dovey (Dyfi). At its unattached end it carries some dunes and a few recurved ends. On its seaward side, near Borth, is a submerged forest, and on its landward side are two distinct features—Cors Fochno, which is a continuation of the submerged forest, and the salt marshes. These marshes have been carefully studied, and may be

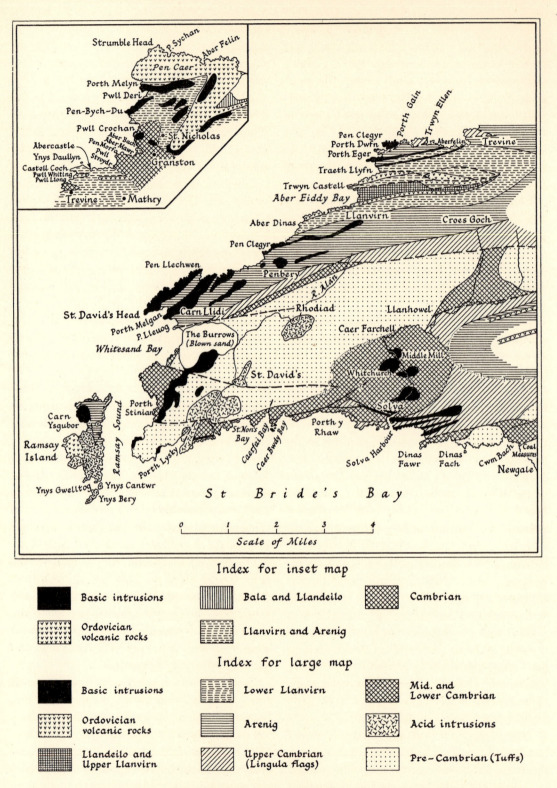

Index for inset map

| | | |
|---|---|---|
| ■ Basic intrusions | ▥ Bala and Llandeilo | ▦ Cambrian |
| Ordovician volcanic rocks | Llanvirn and Arenig | |

Index for large map

| | | |
|---|---|---|
| ■ Basic intrusions | Lower Llanvirn | Mid. and Lower Cambrian |
| Ordovician volcanic rocks | Arenig | Acid intrusions |
| Llandeilo and Upper Llanvirn | Upper Cambrian (Lingula flags) | Pre-Cambrian (Tuffs) |

Fig. 11. Geology of north Pembrokeshire and Strumble Head (inset), after A. H. Cox, J. F. N. Green, O. T. Jones and J. Pringle, *Proc. Geol. Ass., Lond.* **41** (1930).

taken as standard for the west coast. Two main types of vegetation are recognized, a dwarf sward with many channels and salt pans, and a taller variety characterized by *Juncus*. Both rest on silt which is very sandy. Five main plant associations are well defined. The lowest is open and *Salicornia* is dominant; the next highest is mainly a grass sward of *Puccinellia* (*Glyceria*); in the third zone *Armeria* (sea-pink) is dominant, and the fourth zone is again dominated by a grass *Festuca*. These four make up the sward. The fifth and highest zone is the *Juncus* zone. The lawn-like appearance of the marshes, their sandy nature, and the absence of certain plants common in East Anglia, to say nothing of their setting in a mountain-girt estuary, renders them very different in appearance from those on the east coast.

The beautiful valley of the Mawddach is also largely barred by a pebble and sand ridge, Ro Wen, inside which are sand flats and salt marshes. Between the Dovey and the Mawddach there are some cliffs near Fairbourne, but the Dysynni River finds its outlet through masses of shingle.

North of Barmouth, Morfa Dyffryn and Morfa Harlech are extensive sandy forelands with large areas of dune. Morfa Dyffryn has grown to the north and has incorporated Mochras Island, an old moraine, within itself. The dunes, although they do not cover a particularly large area, give the impression of 'wild' country. Between Morfa Dyffryn and Harlech there are some marshes near Pensarn, and the cliffs, now largely protected by the railway, are swathed in boulder-clay. Morfa Harlech is fringed by a cobble beach for about half its length, but since the supply is now cut off by the railway defences, the future of the beach will be interesting. There is no doubt that Morfa Harlech grew northwards, curved round at its unattached end, and finally reached the former island of Llanvihangel-y-Traethau. Harlech Castle was built by Edward I in 1286, and there is good reason for believing that it had then direct connection with the sea, and that Harlech itself was a port. With the continuous northward growth of the spit, it sooner or later became impossible for ships to reach the

port, which consequently decayed. When the embankment between Llanvihangel and the mainland was completed at the beginning of the nineteenth century, the creeks soon deteriorated and the whole area became a marsh.

We must now retrace our steps. In Cardigan Bay there are several sarns (=causeways), of which Sarn Badrig, or St Patrick's Causeway, is far and away the longest. It runs south-westwards from Mochras for about 21 miles, and at a very low ebb perhaps 8 or 9 miles of it may be exposed. It consists of great boulders. Sarn Cynfelin, 2 miles north of Aberystwyth, is about 8 miles long. The others, Sarn y Bwch, Sarn Dewi, and Sarn Cadwgan are shorter. These are natural features, but their precise origin is uncertain. The submerged forest at Borth proves that there has been a downward movement of the land relative to the sea in recent geological times, and there is no doubt that boulder-clay, of which much still covers the cliffs, once extended far out into Cardigan Bay as an extension of the land.

There are many legends of drowned lands in Wales and other Celtic countries, and the whole question of the Sarns has been related to the legend of the lost land of Cantref y Gwaelod. The first written account of this inundation is in the Black Book of Carmarthen (twelfth century). The 'blame' for the flooding is given to one Meredig or Margaret 'who, at times of feasting, allowed the waters of a magic well under her charge to overflow the country'. Popular tradition puts the blame on Seithennin, a renowned drunkard. Tales of this sort grew: in 1662 the connection was made with Sarn Badrig: in the eighteenth and nineteenth centuries myth and legend asserted that Cantref y Gwaelod had disappeared in historic times as a result of carelessness, and it is not surprising that an actual date, A.D. 520, was given as the fatal year. Historic fact in no way whatever supports this. Yet we may have in these tales the germ of truth. Modern opinion regards them as folk tales, that is to say, tales handed down over many generations, and containing in garbled fashion an account of long-forgotten events. The submergence of the forest at Borth almost certainly took place in

131

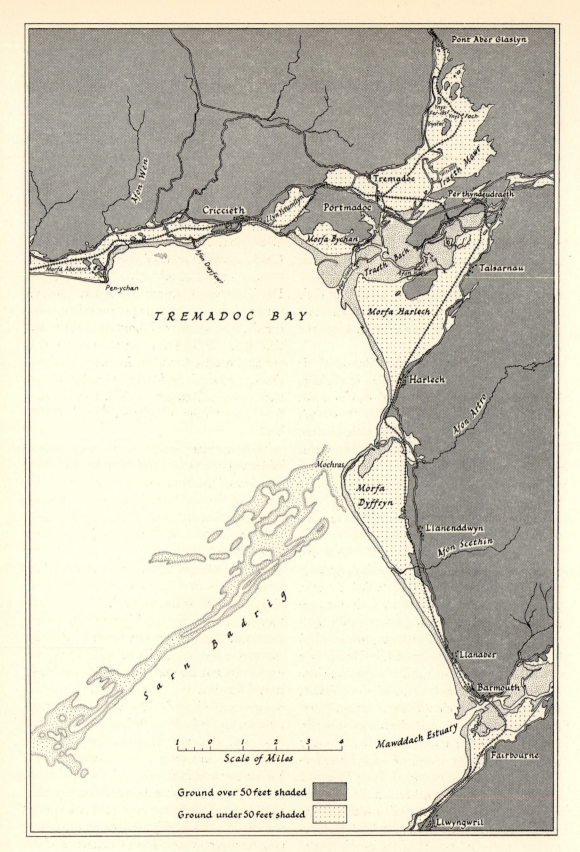

Fig. 12. Coastal features of the northern part of Cardigan Bay.

Neolithic or even in Bronze Age times, and it is therefore quite possible that people did witness it, and that the tales have been handed down from those earlier times and embellished almost out of recognition.

We must now return to the present coast. North of Morfa Harlech are some deep indentations. The first is the Maentwrog Valley, which is separated from the Glaslyn Valley by the Penrhyndeudraeth peninsula. In the Glaslyn Valley, before it was enclosed, the tide reached nearly to Aber Glaslyn bridge, and the valley must have resembled the Mawddach. The idea of enclosing the Glaslyn was first mooted in 1625, but the embankment was not built until 1811. It was breached by a gale in 1812, and finally closed in 1814. Tremadoc and Portmadoc were built after the completion of the embankment. Morfa Bychan is, in one sense, the counterpart of Morfa Harlech. Rising through the morfa are many small 'islands' of solid rock; the Black Rock (Graig Ddu) separates Morfa Bychan from a former deep inlet, now a marsh, Llyn Ystymllyn. The seaward side of the old inlet is now barred by a shingle bank, and shingle becomes increasingly significant along the south coast of the Lleyn peninsula. The various rocky headlands, including that on which Criccieth Castle stands, separate short lines of boulder-clay cliffs which afford abundant coarse beach material. The Afon Dwyfawr is deflected to the east, and the shingle ridge west of Pen ychan, that west of Pwllheli, and those near Llanbedrog and Abersoch are nearly all on the eastern side of the bays in which they rest. Each bay is limited, east and west, by a rocky headland.

St Tudwal's peninsula, formed of sandstones and flags, encloses the fine bay and beach of Porth Caered, and the two small islands are also made of sandstone.

On the west side of St Tudwal's peninsula is Hell's Bay (Porth Neigwl), a wide and dangerous bay largely filled with glacial deposits. The two small islands partly enclosing Aberdaron Bay are parts of a sill of igneous rock, and the bay itself is backed by high cliffs of boulder-clay. Bardsey Island is an isolated mass of pre-Cambrian rocks, and may be compared in general terms with Ramsey Island.

From Aberdaron Bay to Porth Dinllaen the coast is perhaps more interesting than spectacular. It is formed of pre-Cambrian rocks, and glacial drift fills many of the lower parts. There is usually a belt of lower ground adjacent to the shore, but the profile is rather dull. The headlands are often low, but the cliffs rise in height south of Porth Oer. North of Porth Dinllaen— once considered as a possible alternative port to Holyhead—the high ground of Yr Eifl and adjacent headlands is conspicuous, but often spoiled by quarrying. They are formed of igneous rocks. The coast as far as Dinas Dinlle is largely faced with boulder-clay, which forms low cliffs. Locally, as at the Aber Desach, the cliffs give way to a boulder beach and blown sand.

The coast of Anglesey is varied, locally intricate, and interesting. The main drainage of the island is roughly parallel to the Menai Strait, and but a slight subsidence would make another strait between Malldraeth Sands and Red Wharf Bay. The coastal detail between Malldraeth Sands and Holyhead Island shows itself in a series of sandy coves, low rocky headlands, and offshore stacks and shoals. Parts of Holyhead Island are high, and there are fine cliffs in very ancient rocks at the North and South Stacks, and on the main island at Carmel Head. The north coast has several good features, including Cemmaes Bay, Hell's Mouth, Porth Wen Bay, Bull Bay, and Amlwch Harbour. The more sheltered side of the island, Red Wharf Bay to Beaumaris, is less spectacular, but affords the most magnificent views, from, for example, Bwrdd Arthur, of the coast and the Snowdonian mountains.

The Menai Strait marks the line of two former valleys, one flowing north-east and the other south-west. The short middle, and nearly north-south part, just north of Port Dinorwic, was outlined in pre-Glacial times. It is probable that when the ice melted away, the north-east flowing river was for some time dammed by ice which remained in the Irish Sea so that the river was converted into a long lake which overflowed at its head. In this way it cut the short north–south course. Later, submergence converted all the valleys into a marine strait.

The southern entrance to the strait is fringed on the Caernarvon side by the spit of Morfa Dinlle, and on the Anglesey side by the fine dunes of Newborough Warren.

The separation of Holyhead Island is partly the result of streams following lines of faulting, and so etching out valleys which were later drowned.

The remainder of the coast of North Wales is less spectacular, and locally spoiled by inappropriate development. The Great Orme is a fine headland of Carboniferous Limestone and overlooks the Conway Valley. Penmaenmawr and Dwygyfylchi stand on low platforms at the foot of steep mountains. There is a legend of a lost land in Conway Bay. Llys Helig is a mass of seaweed-covered stones visible at low water about a mile off Penmaenmawr. There is a considerable amount of literature referring to it, suggesting that the stones are remains of a former palace. There are also 'maps' showing former 'roads' on what is now the sea floor. There is no basis of truth whatever in these so-called reconstructions. Llys Helig is probably the remains of a moraine, and resembles the scars off the coast of Cumberland.

Beyond Llandudno and Colwyn Bay the shore is fringed by glacial deposits, alluvium, and blown sand. The glacial beds also plaster much of the higher ground, and partly fill up the valleys at Llanddulas, Abergele, and Dyserth.

## THE DEE TO THE SOLWAY

### (PLATES 115–124)

The scenery of this part of our coast is for the most part less attractive than that of most of England and Wales. Much is spoiled by industry or in other ways. Nevertheless there are several features of interest to physiographers.

The inlets of the Dee and Mersey are very unlike one another. The Dee is widely open and narrows inwards. Chester was formerly a port; later Parkgate was used. Now nearly the whole estuary has silted up and salt marsh covers large areas. The Flintshire coast is almost entirely built up, and Flint Castle alone affords a welcome relief in a drab stretch. The Wirral coast of the Dee is less spoiled and often fringed by a boulder-clay cliff. Along the coast between the two estuaries seaside resorts replace industry. Locally hard reefs of Bunter sandstones and pebble beds occur, as at New Brighton, Hilbre Point, and Hilbre Island, However, the coastal formations are mainly post-glacial, and there are extensive dunes at Hoylake, West Kirby, and Wallasey.

The Mersey has a narrow mouth, and widens considerably inland. Both at Liverpool and Birkenhead former inlets have now been reclaimed and built over, or converted into docks. Birkenhead Pool (=Birket or Tranmere Pool) is now the site of the Cammell-Laird works. The former pool of Liverpool covered some 50 acres, but it had disappeared by the middle of the eighteenth century. Some of the wider parts of the Mersey still remain unspoiled, but at and near Runcorn there remains little natural scenery. The narrow mouth of the Mersey induces a strong tidal scour. Thus there is a great difference between the two estuaries—the modern ports of Liverpool and Birkenhead stand in contrast to the former harbours at Chester and Parkgate.

The coast from Liverpool to Southport is low and backed by extensive dunes. Erosion is locally severe, especially near Blundellsands. The tidal range hereabouts is some thirty feet, so that a wide foreshore is exposed at low water. The beach is characterized by broad sand-rolls, three to five feet high, and about 500 feet from crest to crest. Early information about this coast is lacking, and some have held that there were no dunes at Formby before 1690. There seems, indeed, little doubt that by 1750 streets and houses were being buried in sand. The river Alt since about 1698 has flowed southwards across the beach and erosion at Hightown has often been serious. But, as often happens, erosion and accretion alternate with one another; at Formby Point gains up to twelve feet a year have been replaced by losses of nearly twice that amount, and erosion at Hightown has extended to Blundellsands. Erosion is marked when a westerly storm coincides with a high tide.

of the town is the old site of Martin Mere, a shallow basin, the outline of which varied with the rainfall. Available evidence suggests it was formed in the period between the Roman occupation and the Conquest. The present drainage outlet was completed in 1849.

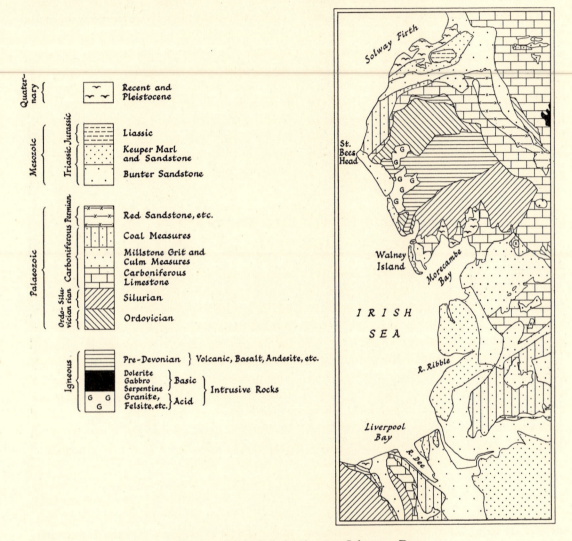

Fig. 13. Geological sketch-map, Solway to Dee.

Southport affords a very different picture. Blown sand is abundant and the coast is extending seawards. In 1736 what is now Lord Street was a salt marsh; by 1834 a new range of dunes had formed and 'Lord Street' was a line of slacks. Later there were extensive reclamations, and the most recent sea bank was made round the municipal golf links in 1931-2. North-east

The channel of the Ribble is subject to change, and much detritus is brought down in floods. Although walls have increased the scour within them, their building has led to much accretion outside, thus producing marshes which are sooner or later inned. On the north bank, dunes formed a natural embankment at Lytham. An interesting feature of these dunes is that plants not native to this country are, or have been, often found in them. The reason is attributed to the seeds and sweepings from American ships which were brought to this district with fodder for the poultry which, before the first world war, was often reared in the dunes.

The Fylde coastline is now almost entirely artificial. Walls front cliffs of boulder-clay, and

Morecambe Bay contains a variety of coastal forms. In the south-east part of it there is an excellent development of salt marsh near Cockerham and Pilling. Samphire is, or was until very recently, cultivated, and much of Cockerham marsh is exploited for turf. The cliffs of the little peninsula on which the remains of Cockersand Abbey stand are protected by a

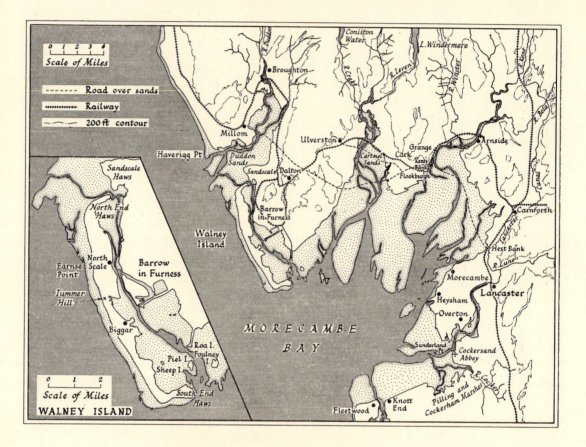

Fig. 14. Morecambe Bay and Walney Island. (Based on Ordnance Survey.)

the groynes show that to the north of Blackpool there is a northerly drift of beach material. There is a break in the defences between South Shore and Lytham St Annes, where there are dunes. To the south of Blackpool beach material travels, in general, towards the Ribble. The northern part of the Fylde coast encloses the estuary of the Wyre, a river on which some interesting research work has recently been carried out by the Hydraulics Research Board. There are salt marshes within the estuary.

wall. Here and around Overton there is a relatively remote district. The outer side of the Heysham peninsula is largely built over, and stands in contrast to the inner side, which slopes gently down to the Lune. Beyond Morecambe the coast is low-lying, and near Hest Bank there are several drumlins into which the sea has cut clear sections. The finest scenery is near Silverdale and Arnside, where there are woods and small valleys, and salt marsh runs right up to the high ground. The estuaries on the north shore are all beautiful, but because of the greater amount of forest, perhaps that of the Leven is the most pleasing. Near Cark there is a good deal of reclaimed marsh. Humphrey Head is a bold headland of Carboniferous Limestone.

Walney Island has a fairly simple outer coast, but an irregular inner one. It is mainly agricultural, but Barrow has spread on to its mid-part. The beaches are shingly. The origin of the island is associated with the Ice Age. It has been suggested that as the ice retreated northwards from Furness, a glacial stream deposited banks of gravel which now form the beds between the lower and upper boulder-clays of Walney. Another view is that the island is of the nature of an esker, a ridge of boulder-clay and pebbles which extended from the base of the melting Duddon glacier at a time when sea-level was rather higher than now. Even if Walney is of glacial origin, it assumes today some features characteristic of a barrier beach. Erosion is serious on its western side. Barrow Island is also formed of boulder-clay capped by sands and gravels. Roa, Foulney, and Sheep islands consist only of shingle.

Erosion in the past has been responsible for the disappearance of certain villages. Herte (Hertye) and Fordebottle in Furness, and Argarmeles and Arnoldsdale (Aynesdale) in West Derby are known to have been lost. In other parts extensive reclamations have been made, sometimes of little value since the reclaimed areas were too sandy. Because of the irregular shape of the coast, there have been frequent 'roads' across the sands. These are dangerous because it is so easy to be caught by the tide, because of the deep channels of the rivers, and because of occasional quicksands. Nevertheless wheeled traffic used to pass regularly from Hest Bank to Kent's Bank, and across the Duddon Sands. Coaches ran regularly up to 1857 when the railway was opened. People with local knowledge still cross the sands, but it is foolhardy to do so without expert guidance.

Perhaps Hodbarrow Point, an outcrop of Carboniferous Limestone, may be taken as the north-western corner of the indented coast of Morecambe Bay and the Duddon estuary. West-

wards therefrom, and near Haverigg Point, there are gravel ridges which have deflected the Whicham Beck. Then follows a line of boulder-clay cliffs, which reach a maximum height of 82 feet at one place. The cliffs are locally ravined and are often grass-covered nearly to sea-level. There is a striking difference between the great mass of Black Combe and the green fields of the boulder-clay strip.

Between Eskmeals and Drigg Point is the joint mouth of the Irt, Mite, and Esk. The Irt is deflected southwards and at one time probably reached the sea opposite Kokoarrah Island. There seems to have been no detailed study of the shingle and sand ridges on either side of the joint mouth, and there is no adequate explanation for the present form of the coast. Northwards of Seascale there are more boulder-clay cliffs which, in one stretch, are landward of the River Ehen, which is deflected southwards by a narrow shingle spit. At St Bees the pleasant sandy beach rests on boulder-clay, and behind it are storm beaches of shingle which have obstructed the Rottington and Pow becks.

St Bees Head is the most conspicuous shore feature, and it also marks a divide in the direction of beach drift. South Head is mainly built of flaggy and false-bedded sandstones. At the main headland there are fine cliffs, often sheer for 200–300 feet. They are, in fact, the only prominent cliffs in the whole length of coast described in this chapter. In Saltom Bay the Brockram and Magnesian Limestones are seen.

The next twelve miles are, scenically, completely ruined by the coalfield. The Coal Measures, sometimes plastered with boulder-clay, form the cliffs, but in several places tip-heaps replace the natural cliffs. Coal mines are often close to or actually on the coast. Locally, as just south of Harrington, there is some open agricultural land. The drift is to the north.

North of Maryport there is abundance of sand and shingle. The raised beach becomes more and more obvious, and from near Workington to Beckfoot it is continuous and forms gently rolling land. The old cliff, cut in boulder-clay, can sometimes be traced inland, and is locally conspicuous in occasional morainic hills—for example, Swarthy Hill, about two miles south of Allonby. There are dunes in Allonby Bay and from Beckfoot to Silloth. On the beach are many scars, the coarse material of former moraines or drumlins.

The northward movement of beach material is shown by the distribution of coal sand from the coalfield farther south. Grune Point is extending, and in Morecambe Bay the area gained by accretion is greater than that lost by erosion. The raised beach and terraces of warp, fronted by salt marshes, prevail all along the Solway coast which is low and flat. The basal rocks are Trias, but are nearly always covered by boulder-clay, warp, and peat. The channels of the estuary and rivers wind about in extensive flats of yellow, mainly quartz, sand. The general setting is beautiful and must be considered in relation to the distant hills and mountains on either side of the estuary.

## NORTHUMBERLAND, DURHAM AND NORTH YORKS
### (PLATES 125–143)

These three parts of the coast may be grouped together for convenience, but each is a distinct unit.

The coast north of the Tweed is formed of cliffs cut in Lower Carboniferous rocks, and is seen to best advantage in Marshall Meadow's Bay. The Tweed valley is deep and south of it the same rocks outcrop on the coast almost as far as Holy Island. They form low cliffs or rocky platforms which sometimes clearly show their geological structure.

The most interesting part of the coast includes Holy Island, the Farnes, Bamburgh, Dunstanburgh, Craster, and Cullernose Point. Holy Island (Lindisfarne) encloses wide sands and flats. The island itself consists of three or more separate islands joined together by shingle beaches. Its size has been increased by wind-blown sand; the Snook is built of dunes 30–40 feet high. Much of the solid part of the island is sandstone, but the castle stands on a dolerite crag which extends to St Cuthbert's Island.

Ross Links is a moorland fringed on the seaward side by a belt of dunes; the inner low-lying part, resting on boulder-clay, is easily flooded. The northern area of the links encloses Fenham Flats; the southern half partly shuts in Budle Bay. At Budle Point the Whin Sill makes its first appearance, and is finely polished and fluted by blown sand. On the south side of the bay the whinstone rises to form a ridge. The Farne Islands are also part of the sill. How many islands there are depends upon whether the observer distinguishes between islets and stacks, or between those parts permanently or occasionally above water-level. The group is divided by Ox Scar Road. They have steep columnar cliffs and flattish tops. Some of the more prominent chasms—for example, St Cuthbert's Gut—follow fissures. In general, the most prominent cliffs face south and west since the sill is gently inclined to the north-east. There are a few patches of sedimentary rocks caught up in the sill.

The sill follows the coast as far south as Cullernose Point, although there are wide interruptions of blown sand, as near Bamburgh where the castle itself stands on the dolerite crags. There are also expanses of sedimentary rocks of Carboniferous age, mainly sandstones, shales, and limestones. These often form rocky platforms, and show differential erosion. Beadnell Bay is dune-fringed; Newton Point projects on account of the hardness of the whinstone and limestone of which it is formed. Farther south sandstones and shales outcrop on the coast, but locally, as at Out Carr and part of Embleton, the Whin reappears.

There are more dunes at Embleton and Newton, and at Dunstanburgh Castle the columnar Whin, overlying limestone, forms a prominent point. The sill is inclined seawards south of the castle, and the foreshore rests on its surface. It rises inland to form a west-facing scarp. It finally leaves the coast at Cullernose Point, just south of which faulting has considerably disturbed the rocks; and several coastal features, including Boulmer Haven and the small bay south of Seaton Point, are partly outlined by faults.

At Alnmouth the links extend in front of the old cliff, and the river, before 1806, flowed round Church Hill, but in that year it broke through a low ridge and so made the present island. Dunes fringe the coast most of the way from the Point, Spital Point, and Hawk's Cliff. The coastal drift is to the south and is exemplified in the small spits which deflect the Aln, Coquet, Lyne, Wansbeck, and Blyth.

The coast from the Coquet to the Tyne is almost entirely industrial. Its natural beauty is disfigured by collieries, railways, and houses. Nowhere is this better seen than at Cambois. The fine beach and dunes give place immediately to railway sidings and industrial squalor. South

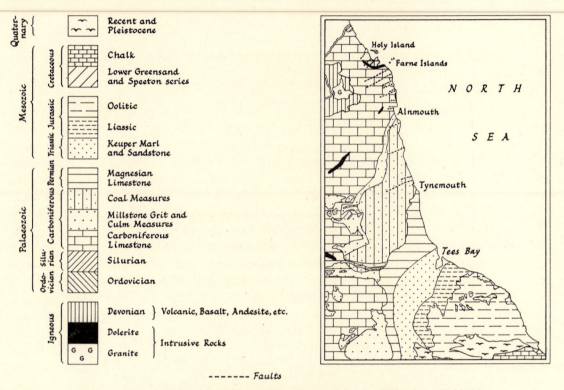

Fig. 15. General geology of the north-east coastal areas. (Trias is undivided.)

Aln to the Coquet, which now reaches the sea at Amble. In pre-Glacial times the mouth was about four miles farther south in Druridge Bay. Between Amble and Sunderland there are several wide-open bays, all of which are sandy. In each of them boulder-clay reaches the sea, and they mark the outlets of pre-Glacial valleys. They are, from north to south, Druridge, the bays north and south of Blyth, Whitley and Whitburn. Between the bays are outcrops of harder rocks usually capped by boulder-clay. Minor inlets are enclosed by projecting promontories of Carboniferous rocks as at Newbiggin of St Mary's Island there is almost continuous town as far as the Tyne. The contrast between the coast north and south of the Coquet results from the fact that the rocks in the southern part are the Productive Coal Measures.

South of the Tyne, which, like the Tweed, has cut down considerably in post-Glacial times, the Coal Measures give place to Permian rocks, which make a coast unique in Britain. The natural cliffs are steep, often vertical, and the stacks in Marsden Bay are fine. Concretions in the Magnesian Limestone can be examined in the stack called the Parson's Rock and in the

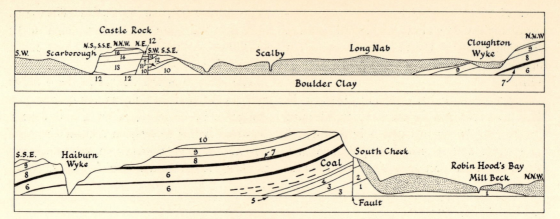

| | | |
|---|---|---|
| | | 19 Upper Calcareous Grit |
| | | 18 Upper Limestone |
| | | 17 Middle Calcareous Grit |
| | Middle Oolite | 16 Lower Limestone |
| | | 15 Passage Beds |
| | | 14 Lower Calcareous Grit |
| | | 13 Oxford Clay |
| | | 12 Kellaways Rock |
| Jurassic | | 11 Cornbrash |
| | | 10 Upper Estuarine Series (moor grit at base) |
| | | 9 Grey Limestone Series (marine) |
| | Lower Oolite | 8 Middle Estuarine Series |
| | | 7 Millepore Bed (marine) |
| | | 6 Lower Estuarine Series (with Eller Beck Bed, marine) |
| | | 5 The Dogger (marine) |
| | | 4 Blea Wyke Beds (sandy) |
| | | 3 Upper Lias (with Jet Rock and Alum Shale) |
| | | (The *A. jurensis* zone is near the top of this bed) |
| | Lias | 2 Middle Lias (with ironstones) |
| | | (including *A. capricornus* zone) |
| | | 1 Lower Lias (including *A. jamesoni* and *A. oxynotus* zones) |

Fig. 16. Cliff section of the Yorkshire coast from Scarborough to Robin Hood's Bay, after
J. F. Blake, *Proc. Geol. Ass., Lond.* **12** (1891).

Roker cliffs, and the formation of caverns at Hole Rock near Ryhope. Between Seaham and Hartlepool there are several deeply cut valleys, called denes, for example, at Horden, Ryhope, Seaham, Hawthorn, Foxhole, Castle Eden, Hesleden and Crimden. They are mainly excavated in boulder-clay, but some also cut well down into the underlying limestone.

The natural coast of Durham was beautiful; that near Hawthorn Dene is perhaps the best remaining part. Locally colliery waste is dumped below high-water mark, and at Easington Colliery the cliff-face in part is colliery waste tipped over the natural cliff. Near Hartlepool the coast changes. The town itself rests on an outlying mass of limestone—The Heugh—joined to the mainland by a spit. As far as the Tees marshes, the cliffs are cut in sandstone of Triassic age. In the mouth of the Tees there has been a great deal of reclamation. At the close of the Ice Age the Tees and various small streams ran into an estuary. The southward travel of beach material (so well seen by the distribution of the coal sand) and the fine stuff carried by the tide were largely responsible for filling the estuary with sediment, and led to the formation of mud-flats, locally known as slems. Even in 1920 the natural marsh at Greatham was but a fragment of its original self; there has been more change since then. There are extensive flats on the Yorkshire side, where industrialization has also wrought great changes.

At Redcar and Saltburn the cliffs are in boulder-clay and there is a beach. East of Saltburn the Lias cliffs are prominent. Hunt Cliff, despite the softness of the Lower Lias shales, stands out in contrast to the even less resistant boulder-clay. Between Skinningrove and Whitby several small becks reach the coast. Some are in their original valleys, some are in courses imposed on them by the plugging of their former valleys by boulder-clay, and the two at Sandsend replace a single stream, and are busily employed in removing the boulder-clay which filled the original valley. The Esk at Whitby is gorge-like; it is of pre-Glacial age and follows a line of faulting.

South of Whitby the coast is fine and mainly unspoiled. Fig. 16 shows its general nature and emphasizes the abundance of boulder-clay in some of the bays. Robin Hood's Bay is beautiful, especially as seen from Ravenscar. It is backed by the open moorland of Fylingdales. The Peak cliffs at the south end reach about 600 feet, and south therefrom there is an undercliff, and at Hayburn Wyke a shingly beach and a deep wooded ravine. Cloughton Wyke is similar but smaller. From Peak to Scarborough the coast runs to the south-south-east and may have been determined in part by faulting. At the north

bay at Scarborough the cliffs are in drift, and the promontory itself is formed of sandstone.

Cliffs continue south from Scarborough. The rocks are Jurassic and locally affected by faulting. The notch left by the fault in Osgodby Nab gives access to Cayton Bay, which is almost entirely the product of faulting and contains a good beach shut in by picturesque headlands. Carr Naze and Filey Brigg are conspicuous. The Brigg is made of grit, and the rocks of the promontory are worn by erosion into steps and holes. The Doodles are cove-like features. Beyond the Brigg the cliffs are mainly composed of boulder-clay, which is tenacious and hard, so that the cliffs are more or less vertical, and often furrowed or ravined. The Speeton Clay (Lower Cretaceous) follows to the south, but in Filey Bay it is much slipped and confused. Between Bempton and Flamborough the hard, thick, Chalk-with-flints forms a magnificent range of almost vertical cliffs, with perforations and arches, ridges and gullies, especially near Thornwick. Flamborough Head is interesting, but spoiled by unsightly development. The arch near the North Sea landing in cut along a joint or fault; the King and Queen rocks and Adam and Eve pinnacles are stacks; there are good arches at Kindle Scar; Pigeon Hole is a blow-hole. All the headland is drift-covered, and the cliff profile is more or less vertical in the Chalk, but sloping in the overlying drift.

The Jurassic and Cretaceous cliffs of Yorkshire are among the finest in this country. Their setting in agricultural land, and their height and sheer faces give them great character. They are comparable with those in Dorset, but lack the interesting detail of those on the south coast. Nevertheless, Staithes and Fylingdales Moor have no counterparts in Dorset.

## FLAMBOROUGH HEAD TO HUNSTANTON

### (PLATES 142–150)

Nearly all of this long stretch of coast is a recent gift to the country. Holderness and much of north and east Lincolnshire are built of glacial deposits, and all around the Wash, including the Fenland, the alluvial deposits are of recent geological age. The former coastline runs approximately along the line of the Yorkshire and Lincolnshire Wolds from Flamborough Head to Spilsby, although this is quite an arbitrary southern limit. Fenland deposits, silt near the sea and peat inland, extend almost as far as Lincoln, and the coast of the former Fenland gulf coincided roughly with a line joining Firsby, Tattershall, Lincoln, Billinghay, Peterborough, Earith, Cottenham, Burwell, Denver, and King's Lynn. But there were also fenlands in Yorkshire; long areas of swamp extended from the Humber up the Ouse, Don, Idle, Trent, and Ancholme valleys. It will be appreciated that a coast formed of glacial and alluvial materials is one likely to suffer a good deal of change. The alluvial deposits imply deposition in quiet waters, and the gradual formation of new land. The process can be watched in the Humber and the Wash today. The glacial deposits, where they face the sea as cliffs, are subject to erosion, locally serious.

Along the coast of Holderness there are low or medium-high cliffs. At places, for example, Hornsea and Withernsea, they are protected, but elsewhere erosion cuts them away at the rate of perhaps five or six or more feet a year. This means that the protected areas are gradually becoming 'capes', and the necessity for defence against erosion increases. Many villages have disappeared along this coast, and it is reliably estimated that since Roman times more than 80 square miles of land have been washed away. As a general rule erosion is greatest where the cliffs are lowest. There is a sandy beach at the foot of the cliffs, and the larger stones and boulders from the drift go to form the stone banks a little way offshore.

Partly counterbalancing this loss is a consider-able amount of accretion in the Humber, one of our muddiest rivers. It is now generally agreed that the mud is mainly carried in by the tide, and both by natural means and by warping (a prac-tice perhaps more common in the past) large additions of fertile lands have been made, especially at Sunk Island and Cherry Cob sands. Spurn Head is also fed by the waste from Holderness. Its maintenance has of recent years depended much upon groynes and the abundance of marram grass. Its form repre-sents a balance between the influence of the North Sea and that of the river. It has fluctuated a great deal in recent centuries, and some of its history can be traced in the several lighthouses which have been built upon it. It is also not improbable that it occasionally grows to such a length that the distal part becomes separated, and the sand forming it is driven on to the Lincolnshire coast where, at Donna Nook, there are vast expanses of sand, the origin of which may reasonably be explained in this way.

On the Lincolnshire coast there is now no 'live' cliff; that at Cleethorpes (the only cliff between Holderness and Hunstanton) is protec-ted by the promenade. All the coast of Lincoln-shire is flat; in the north there are extensive sand-dunes. Nowhere in Britain suffered so severely as mid-Lincolnshire in the great storm and floods of 1953. North of Mablethorpe the dunes are wide and beach profile flat; south of that place the beach is narrower, and the dunes smaller or absent. This prevails to near Skegness. From there southwards the beach is wider and there are more dunes. Now-adays the coast from, approximately, Mable-thorpe to near Skegness is entirely protected. It took four or five years for the beach to re-plenish itself after the storm of 1953, during which large stretches were swept completely clear of sand, and the underlying boulder-clay severely eroded. A casual observer on this coast before the flood might well have concluded that there was little or no erosion; those who knew it

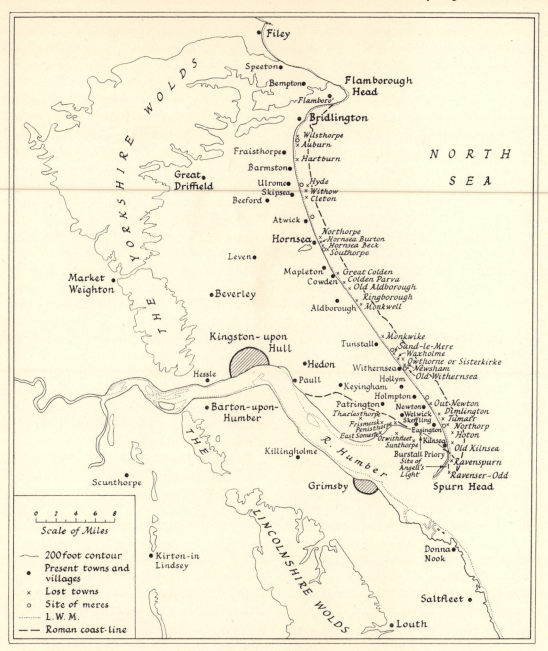

Fig. 17. The lost towns of East Yorkshire and Holderness 'Bay', after T. Sheppard.

well realized that from time to time storms swept away the sand and cut into the clay, but it was not until the great storm in 1953 that it was sufficiently appreciated how serious this damage could be.

One of the most interesting localities on the Lincolnshire coast is Ingoldmells Point. There it has been demonstrated, partly on the basis of plant remains, partly by traces of Roman occupation which are now below mid-tide level, that the coast has been slowly sinking over many centuries.

The Wash is the unfilled remnant of the former Fenland gulf. All round its shores reclamations are made from time to time. Since Roman times it has been estimated that some

8000 acres have been gained on the Norfolk coast, and 37,000 acres in South Holland. The longer natural accretion takes place before embanking the better. The foreshore outside a new bank usually becomes grass marsh in about ten years, but it is best to let another twenty-five or thirty years elapse before inning it. There are numerous sea-walls around the Wash, and it is often noticeable that if one

valleys in the Chalk to a more distant North Sea. They were enlarged both by erosion and subsidence. Eventually, the Chalk disappeared from the gaps and the sea was brought directly on to the underlying rocks.

The shores of the Wash today are for the most part fringed by wide belts of marsh, only broken where the main rivers flow out. On the Norfolk coast there is a good deal of sand and

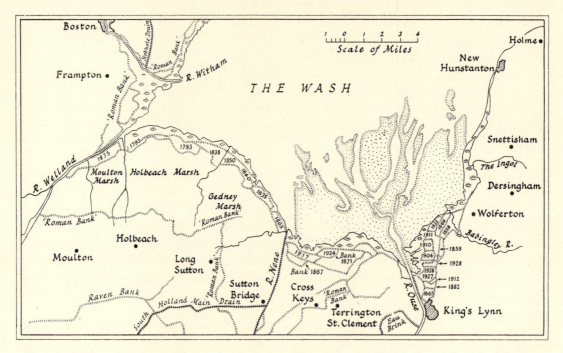

Fig. 18. Reclamations in the Wash, after O. Borer, 'Changes in the Wash', *Geogr. J.* **93** (1939), 491.

goes inland from the Wash the level of the marshes decreases.

The Humber and the Wash, although unlike one another in appearance, have a similar origin. Both inlets breach the Chalk, which at one time was continuous from Norfolk to Yorkshire. Before the breaches were made the rivers flowed, like those of the Weald today, through narrow

shingle north of Snettisham, and the Heacham river has been deflected southwards. The White Chalk, Red Chalk, and Lower Greensand appear at Hunstanton to form one of the best known and highly coloured lines of cliff in Britain. Traces of older cliffs can be followed farther south, and are conspicuous near the road and railway a little north of Wolferton station.

## HUNSTANTON TO THE THAMES

### (PLATES 151–167)

This length of coast may be conveniently subdivided into three parts—the marshland of north Norfolk, the alternating lines of soft, often glacial, cliffs and river mouths from Wey-

bourne to Clacton, and the remainder of the coast of Essex. It is throughout a coast along which great changes have taken place, both of accretion and erosion. It is also one along which it is often possible to correlate physiographical evolution with historic events. The land within is low, and where cliffs reach the sea they are soft and liable to erosion.

The marshland begins immediately east of Hunstanton. The coloured beds disappear, and a line of old cliffs runs inside the marshes, and shows as a gentle inclination landwards from the main coast road. The sea shallows gradually, and wide expanses of sand are exposed at low-water. The waves separate sand and shingle, and where the latter is abundant it is usually high up on the beach, and often forms ridges and bars, behind some of which there are extensive salt marshes. All stages of growth of these ridges can be traced from miniature ones such as occur off Stiffkey, through the more developed examples at Thornham and Wells, and finally to such major features as Scolt Head Island and Blakeney Point. The ridges may join the main-land as at Holme and Blakeney, elsewhere they are of the nature of barrier beaches, of which Scolt Head Island is the best example.

The main ridges of shingle often throw off a number of lateral ridges, each of which was once the end of the main island or spit of which it now forms a part. At Blakeney there is abundant shingle, and the main ridge is on a far bigger scale than elsewhere on this coast, and the laterals are in groups. At Scolt Head Island the laterals are rather more widely spaced and longer. Like those at Blakeney they carry dunes on them. In places the dunes are simple forms, and every stage exists from a tiny heap of sand, often beginning to grow around a tuft of marram grass, to complex groups like those on Hut Hills and House Hills at Scolt, and on the headland at Blakeney.

Inside the shingle and dune ridges accretion goes on and salt marshes develop. The Norfolk marshes are more colourful than those on the west and south coasts, and are usually built up of much firmer and thicker mud. They increase in height, especially once a plant cover has formed, since this helps to trap the tidal silt and so leads to deposition and upward growth. The slightly different levels of the marshes can be appreciated if a big flooding tide is watched from a high

dune—it will then be clear that the newer marshes are first covered. If, later, at low-water a traverse is made over the marshes, the amount of accretion in each, the plant cover and plant types, and the development of creeks and pools will be seen to fall into a well-marked sequence. Some of the older marshes have been embanked; those at Holkham, Wells, and Holme, are examples. The sand-hills enclosing the woods at Holkham rest partly on shingle-ridges, which are of similar origin to Scolt Head Island. The trees were planted about the middle of the last century.

The cliffs between Weybourne and Happis-burgh are glacial. A walk along the beach reveals how quickly they vary in composition and how, near Sheringham and Cromer, the stratification is confused and contorted. They are liable to severe erosion, particularly where silt and incoherent sands and gravels occur. In recent years Overstrand has suffered greatly, and Mundesley may be compared with Hornsea and Withernsea in Holderness—the cliffs on either side are constantly wearing away so that the town with its sea-defences is becoming increasingly vulnerable. The cliffs vary in height, from peaks like Skelding and Beeston Hills at Sheringham to the low and protected cliffs at Walcott and Bacton.

Beyond Happisburgh the cliffs soon disappear, and for some miles the beach is backed by a single line of dunes behind which the country is low and falls to Broadland. Since the floods of 1938 and 1953 extensive engineering works have been carried out, and sea-walls, often dune-covered, protect the coast. The beach is steeper than in north Norfolk, and is characterized by long gullies. In the 1953 storm a serious breach was made at Palling where now there is a pull-over to reach the beach. A curious change occurs at Winterton; erosion gives place to accretion, and a ness has formed. To the north-west of the village there are extensive dunes, and to the south the old cliffs are fronted by a hollow, on the sea-ward side of which there are long ridges of dunes. Erosion begins again at Hemsby, and between that place and Caister it is often serious.

There was formerly a wide mouth of the Bure and Yare south of Caister, and the sea penetrated the valleys in which the Broads now lie. Later a sand bank, on which Yarmouth stands, was formed and there were two mouths, one on the north which soon silted up, and one on the south which still remains. This has been subject to many changes, and at one time was almost opposite to Corton. Where there are no defences erosion is serious south of Yarmouth Haven, the piers of which hold up material on their north side and cause the constant accumulation of beach at Yarmouth. Lowestoft Ness projects in front of old cliffs, and south of the harbour erosion is again a problem, apart from a short stretch a little north of Kessingland. The short lines of cliff on the Suffolk coast are all soft, of moderate height, and subject to erosion. The gaps, originally the mouths of small streams, are now dammed by sand and shingle beaches. At Southwold and Dunwich great changes have taken place, and throughout the centuries much land has been lost at Dunwich, which in medieval times was important both ecclesiastically and as a port.

All the way along this coast beach material travels southwards. This has been responsible for the blocking of small streams, the deflection of the Yare, and especially of the Alde which is turned so that it flows for ten or eleven miles parallel to the coast. The deflecting spit, Orford Ness, is made of many ridges of shingle, and when Orford Castle was built in 1262 the spit had grown to a place marked on large-scale maps as Stony Ditch Point. It remained in this position for some considerable time, and thus was a protection to Orford; later it lengthened and Orford decayed as a port. There is a small spit at the mouth of the Deben, and Landguard Point is prominent at the joint mouth of the Orwell and Stour.

One of the great charms of this coast lies in the tidal valleys of the Blyth, Alde, Deben, Orwell, Stour, and the Essex rivers. They are the result of submergence of a low coast, and since the downward movement is slow enough for marsh growth to keep pace with it, they locally contain wide expanses of saltings.

The peninsula between the Stour and the Blackwater resembles in broad outline the east Suffolk coast. Hamford Water is a ramifying inlet with much salt marsh, and at Harwich, Dovercourt and Walton there are low cliffs of London Clay, capped by Red Crag; the London Clay also crops out at Clacton. Shingle drifts towards the Colne estuary, and around Mersea Island, also formed of London Clay, this clay and tidal mud form large expanses.

A Roman, as well as a modern nuclear, station stands at the mouth of the Blackwater at Bradwell (Othona). The study of the coast of this part of Essex resolves itself into a study of rivers and embanking; in the past there have been several changes of river courses, but their history involves the late Tertiary evolution of the Thames. Salt marshes fringe the coast from Bradwell to the Crouch. Foulness and adjacent islands are low-lying and in the nature of delta deposits of the Thames. Several creeks between them have been dammed, and the islands thus increased in size. The mud is thick, 80 feet on the north of the Crouch, and 30–40 feet in the Foulness creeks—clear evidence of a time when the land stood at a higher level.

Precisely where our study of the coast ends and that of the Thames begins is debatable; but Canvey Island and the neighbourhood of Southend are not only interesting from the point of view of saltmarsh development and embanking, but also because archaeological excavations at Southchurch have revealed evidence of post-Roman subsidence. In fact, the Thames is but another drowned mouth, but on a far larger scale than those in Essex and Suffolk.